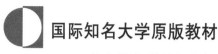

国际知名大学原版教材
—— 信息技术学科与电气工程学科系列

Lessons from AlphaZ
Model Predictive, and Adaptive Control

阿尔法零对最优模型预测自适应控制的启示

[美] 德梅萃·P. 博塞克斯　著
(Dimitri P. Bertsekas)

清华大学出版社
北京

北京市版权局著作权合同登记号　图字：01-2024-2268

Original English Language Edition: Lessons from AlphaZero for Optimal, Model Predictive, and Adaptive Control by Dimitri P. Bertsekas.
Copyright © Dimitri P. Bertsekas, 2024.
All rights reserved.
Athena Scientific, Belmont, MA, USA.

版权所有，侵权必究。举报：010-62782989，beiqinquan@tup.tsinghua.edu.cn。

图书在版编目（CIP）数据

阿尔法零对最优模型预测自适应控制的启示 = Lessons from AlphaZero for Optimal, Model Predictive, and Adaptive Control：英文 ／ （美）德梅萃·P.博塞克斯（Dimitri P. Bertsekas）著. -- 北京：清华大学出版社，2025.3. -- （国际知名大学原版教材）. -- ISBN 978-7-302-68471-8

Ⅰ. TP181
中国国家版本馆 CIP 数据核字第 2025BS6678 号

责任编辑：古　雪
封面设计：傅瑞学
责任校对：李建庄
责任印制：丛怀宇

出版发行：清华大学出版社
　　　　网　　址：https://www.tup.com.cn，https://www.wqxuetang.com
　　　　地　　址：北京清华大学学研大厦A座　　邮　编：100084
　　　　社 总 机：010-83470000　　　　　　　　邮　购：010-62786544
　　　　投稿与读者服务：010-62776969，c-service@tup.tsinghua.edu.cn
　　　　质量反馈：010-62772015，zhiliang@tup.tsinghua.edu.cn
　　　　课件下载：https://www.tup.com.cn,010-83470236
印 装 者：三河市君旺印务有限公司
经　　销：全国新华书店
开　　本：155mm×235mm　　印张：15.5　　字　数：347 千字
版　　次：2025 年 4 月第 1 版　　　　　　　　印　次：2025 年 4 月第 1 次印刷
印　　数：1～1500
定　　价：79.00 元

产品编号：097168-01

出 版 说 明

　　本书为英文影印版,书中所有的数学符号、公式、参考文献格式等都遵照原书,未作更改。特此说明。

清华大学出版社

ABOUT THE AUTHOR

Dimitri Bertsekas studied Mechanical and Electrical Engineering at the National Technical University of Athens, Greece, and obtained his Ph.D. in system science from the Massachusetts Institute of Technology. He has held faculty positions with the Engineering-Economic Systems Department, Stanford University, and the Electrical Engineering Department of the University of Illinois, Urbana. Since 1979 he has been teaching at the Electrical Engineering and Computer Science Department of the Massachusetts Institute of Technology (M.I.T.), where he is McAfee Professor of Engineering. In 2019, he joined the School of Computing and Augmented Intelligence at the Arizona State University, Tempe, AZ, as Fulton Professor of Computational Decision Making.

Professor Bertsekas' teaching and research have spanned several fields, including deterministic optimization, dynamic programming and stochastic control, large-scale and distributed computation, artificial intelligence, and data communication networks. He has authored or coauthored numerous research papers and nineteen books, several of which are currently used as textbooks in MIT classes, including "Dynamic Programming and Optimal Control," "Data Networks," "Introduction to Probability," and "Nonlinear Programming." At ASU, he has been focusing in teaching and research in reinforcement learning, and he has developed several textbooks and research monographs in this field since 2019.

Professor Bertsekas was awarded the INFORMS 1997 Prize for Research Excellence in the Interface Between Operations Research and Computer Science for his book "Neuro-Dynamic Programming" (co-authored with John Tsitsiklis), the 2001 AACC John R. Ragazzini Education Award, the 2009 INFORMS Expository Writing Award, the 2014 AACC Richard Bellman Heritage Award, the 2014 INFORMS Khachiyan Prize for Life-Time Accomplishments in Optimization, the 2015 MOS/SIAM George B. Dantzig Prize, and the 2022 IEEE Control Systems Award. In 2018 he shared with his coauthor, John Tsitsiklis, the 2018 INFORMS John von Neumann Theory Prize for the contributions of the research monographs "Parallel and Distributed Computation" and "Neuro-Dynamic Programming." Professor Bertsekas was elected in 2001 to the United States National Academy of Engineering for "pioneering contributions to fundamental research, practice and education of optimization/control theory, and especially its application to data communication networks."

ATHENA SCIENTIFIC
OPTIMIZATION AND COMPUTATION SERIES

1. Lessons from AlphaZero for Optimal, Model Predictive, and Adaptive Control by Dimitri P. Bertsekas, 2022, ISBN 978-1-886529-17-5, 245 pages

2. Abstract Dynamic Programming, 3rd Edition, by Dimitri P. Bertsekas, 2022, ISBN 978-1-886529-47-2, 420 pages

3. Rollout, Policy Iteration, and Distributed Reinforcement Learning, by Dimitri P. Bertsekas, 2020, ISBN 978-1-886529-07-6, 480 pages

4. Reinforcement Learning and Optimal Control, by Dimitri P. Bertsekas, 2019, ISBN 978-1-886529-39-7, 388 pages

5. Dynamic Programming and Optimal Control, Two-Volume Set, by Dimitri P. Bertsekas, 2017, ISBN 1-886529-08-6, 1270 pages

6. Nonlinear Programming, 3rd Edition, by Dimitri P. Bertsekas, 2016, ISBN 1-886529-05-1, 880 pages

7. Convex Optimization Algorithms, by Dimitri P. Bertsekas, 2015, ISBN 978-1-886529-28-1, 576 pages

8. Convex Optimization Theory, by Dimitri P. Bertsekas, 2009, ISBN 978-1-886529-31-1, 256 pages

9. Introduction to Probability, 2nd Edition, by Dimitri P. Bertsekas and John N. Tsitsiklis, 2008, ISBN 978-1-886529-23-6, 544 pages

10. Convex Analysis and Optimization, by Dimitri P. Bertsekas, Angelia Nedić, and Asuman E. Ozdaglar, 2003, ISBN 1-886529-45-0, 560 pages

11. Network Optimization: Continuous and Discrete Models, by Dimitri P. Bertsekas, 1998, ISBN 1-886529-02-7, 608 pages

12. Network Flows and Monotropic Optimization, by R. Tyrrell Rockafellar, 1998, ISBN 1-886529-06-X, 634 pages

13. Introduction to Linear Optimization, by Dimitris Bertsimas and John N. Tsitsiklis, 1997, ISBN 1-886529-19-1, 608 pages

14. Parallel and Distributed Computation: Numerical Methods, by Dimitri P. Bertsekas and John N. Tsitsiklis, 1997, ISBN 1-886529-01-9, 718 pages

15. Neuro-Dynamic Programming, by Dimitri P. Bertsekas and John N. Tsitsiklis, 1996, ISBN 1-886529-10-8, 512 pages

16. Constrained Optimization and Lagrange Multiplier Methods, by Dimitri P. Bertsekas, 1996, ISBN 1-886529-04-3, 410 pages

17. Stochastic Optimal Control: The Discrete-Time Case, by Dimitri P. Bertsekas and Steven E. Shreve, 1996, ISBN 1-886529-03-5, 330 pages

Contents

1. AlphaZero, Off-Line Training, and On-Line Play
- 1.1. Off-Line Training and Policy Iteration . . . . p. 3
- 1.2. On-Line Play and Approximation in Value Space - Truncated Rollout . . . . p. 6
- 1.3. The Lessons of AlphaZero . . . . p. 8
- 1.4. A New Conceptual Framework for Reinforcement Learning p. 11
- 1.5. Notes and Sources . . . . p. 14

2. Deterministic and Stochastic Dynamic Programming
- 2.1. Optimal Control Over an Infinite Horizon . . . . p. 20
- 2.2. Approximation in Value Space . . . . p. 25
- 2.3. Notes and Sources . . . . p. 30

3. An Abstract View of Reinforcement Learning
- 3.1. Bellman Operators . . . . p. 32
- 3.2. Approximation in Value Space and Newton's Method . . p. 39
- 3.3. Region of Stability . . . . p. 46
- 3.4. Policy Iteration, Rollout, and Newton's Method . . . . . p. 50
- 3.5. How Sensitive is On-Line Play to the Off-Line Training Process? . . . . p. 58
- 3.6. Why Not Just Train a Policy Network and Use it Without On-Line Play? . . . . p. 60
- 3.7. Multiagent Problems and Multiagent Rollout . . . . . . p. 61
- 3.8. On-Line Simplified Policy Iteration . . . . p. 66
- 3.9. Exceptional Cases . . . . p. 72
- 3.10. Notes and Sources . . . . p. 79

4. The Linear Quadratic Case - Illustrations
- 4.1. Optimal Solution . . . . p. 82
- 4.2. Cost Functions of Stable Linear Policies . . . . p. 83
- 4.3. Value Iteration . . . . p. 86

- 4.4. One-Step and Multistep Lookahead - Newton Step
  Interpretations . . . . p. 86
- 4.5. Sensitivity Issues . . . . p. 91
- 4.6. Rollout and Policy Iteration . . . . p. 94
- 4.7. Truncated Rollout - Length of Lookahead Issues . . . . p. 97
- 4.8. Exceptional Behavior in Linear Quadratic Problems . . . p. 99
- 4.9. Notes and Sources . . . . p. 100

5. Adaptive and Model Predictive Control

- 5.1. Systems with Unknown Parameters - Robust and
  PID Control . . . . p. 102
- 5.2. Approximation in Value Space, Rollout, and Adaptive
  Control . . . . p. 105
- 5.3. Approximation in Value Space, Rollout, and Model
  Predictive Control . . . . p. 109
- 5.4. Terminal Cost Approximation - Stability Issues . . . . p. 112
- 5.5. Notes and Sources . . . . p. 118

6. Finite Horizon Deterministic Problems - Discrete Optimization

- 6.1. Deterministic Discrete Spaces Finite Horizon Problems . p. 120
- 6.2. General Discrete Optimization Problems . . . . p. 125
- 6.3. Approximation in Value Space . . . . p. 128
- 6.4. Rollout Algorithms for Discrete Optimization . . . . p. 132
- 6.5. Rollout and Approximation in Value Space with Multistep
  Lookahead . . . . p. 149
  - 6.5.1. Simplified Multistep Rollout - Double Rollout . . . p. 150
  - 6.5.2. Incremental Rollout for Multistep Approximation in
    Value Space . . . . p. 153
- 6.6. Constrained Forms of Rollout Algorithms . . . . p. 159
- 6.7. Adaptive Control by Rollout with a POMDP Formulation p. 173
- 6.8. Rollout for Minimax Control . . . . p. 182
- 6.9. Small Stage Costs and Long Horizon - Continuous-Time
  Rollout . . . . p. 190
- 6.10. Epilogue . . . . p. 197

Appendix A: Newton's Method and Error Bounds

- A.1. Newton's Method for Differentiable Fixed
  Point Problems . . . . p. 202
- A.2. Newton's Method Without Differentiability of the
  Bellman Operator . . . . p. 207

A.3. Local and Global Error Bounds for Approximation in
        Value Space . . . . p. 210
   A.4. Local and Global Error Bounds for Approximate
        Policy Iteration . . . . p. 212

References . . . . p. 217

Preface

> With four parameters I can fit an elephant, and with five I can make him wiggle his trunk.†
>
> **John von Neumann**

The purpose of this monograph is to propose and develop a new conceptual framework for approximate Dynamic Programming (DP) and Reinforcement Learning (RL). This framework centers around two algorithms, which are designed largely independently of each other and operate in synergy through the powerful mechanism of Newton's method. We call these the *off-line training* and the *on-line play* algorithms; the names are borrowed from some of the major successes of RL involving games. Primary examples are the recent (2017) AlphaZero program (which plays chess), and the similarly structured and earlier (1990s) TD-Gammon program (which plays backgammon). In these game contexts, the off-line training algorithm is the method used to teach the program how to evaluate positions and to generate good moves at any given position, while the on-line play algorithm is the method used to play in real time against human or computer opponents.

---

† From the meeting of Freeman Dyson and Enrico Fermi (p. 273 of the Segre and Hoerlin biography of Fermi, The Pope of Physics, Picador, 2017): "When Dyson met with him in 1953, Fermi welcomed him politely, but he quickly put aside the graphs he was being shown indicating agreement between theory and experiment. His verdict, as Dyson remembered, was "There are two ways of doing calculations in theoretical physics. One way, and this is the way I prefer, is to have a clear physical picture of the process you are calculating. The other way is to have a precise and self-consistent mathematical formalism. You have neither." When a stunned Dyson tried to counter by emphasizing the agreement between experiment and the calculations, Fermi asked him how many free parameters he had used to obtain the fit. Smiling after being told "Four," Fermi remarked, "I remember my old friend Johnny von Neumann used to say, with four parameters I can fit an elephant, and with five I can make him wiggle his trunk." See also the paper by Mayer, Khairy, and Howard [MKH10], which provides a verification of the von Neumann quotation.

Both AlphaZero and TD-Gammon were trained off-line extensively using neural networks and an approximate version of the fundamental DP algorithm of policy iteration. Yet the AlphaZero player that was obtained off-line is not used directly during on-line play (it is too inaccurate due to approximation errors that are inherent in off-line neural network training). Instead a separate on-line player is used to select moves, based on multistep lookahead minimization and a terminal position evaluator that was trained using experience with the off-line player. The on-line player performs a form of policy improvement, which is not degraded by neural network approximations. As a result, it greatly improves the performance of the off-line player.

Similarly, TD-Gammon performs on-line a policy improvement step using one-step or two-step lookahead minimization, which is not degraded by neural network approximations. To this end it uses an off-line neural network-trained terminal position evaluator, and importantly it also extends its on-line lookahead by rollout (simulation with the one-step lookahead player that is based on the position evaluator).

Thus in summary:

(a) The on-line player of AlphaZero plays much better than its extensively trained off-line player. This is due to the beneficial effect of exact policy improvement with long lookahead minimization, which corrects for the inevitable imperfections of the neural network-trained off-line player, and position evaluator/terminal cost approximation.

(b) The TD-Gammon player that uses long rollout plays much better than TD-Gammon without rollout. This is due to the beneficial effect of the rollout, which serves as a substitute for long lookahead minimization.

An important lesson from AlphaZero and TD-Gammon is that the performance of an off-line trained policy can be greatly improved by on-line approximation in value space, with long lookahead (involving minimization or rollout with the off-line policy, or both), and terminal cost approximation that is obtained off-line. This performance enhancement is often dramatic and is due to a simple fact, which is couched on algorithmic mathematics and is the focal point of this work:

(a) *Approximation in value space with one-step lookahead minimization amounts to a step of Newton's method for solving Bellman's equation.*

(b) *The starting point for the Newton step is based on the results of off-line training, and may be enhanced by longer lookahead minimization and on-line rollout.*

Indeed the major determinant of the quality of the on-line policy is the Newton step that is performed on-line, while off-line training plays a secondary role by comparison.

Significantly, the synergy between off-line training and on-line play also underlies Model Predictive Control (MPC), a major control system design methodology that has been extensively developed since the 1980s. This synergy can be understood in terms of abstract models of infinite horizon DP and simple geometrical constructions, and helps to explain the all-important stability issues within the MPC context.

An additional benefit of policy improvement by approximation in value space, not observed in the context of games (which have stable rules and environment), is that it works well with changing problem parameters and on-line replanning, similar to indirect adaptive control. Here the Bellman equation is perturbed due to the parameter changes, but approximation in value space still operates as a Newton step. An essential requirement within this context is that a system model is estimated on-line through some identification method, and is used during the one-step or multistep lookahead minimization process.

In this monograph we will aim to provide insights (often based on visualization), which explain the beneficial effects of on-line decision making on top of off-line training. In the process, we will bring out the strong connections between the artificial intelligence view of RL, and the control theory views of MPC and adaptive control. Moreover, we will show that in addition to MPC and adaptive control, our conceptual framework can be effectively integrated with other important methodologies such as multiagent systems and decentralized control, discrete and Bayesian optimization, and heuristic algorithms for discrete optimization.

One of our principal aims is to show, through the algorithmic ideas of Newton's method and the unifying principles of abstract DP, that the AlphaZero/TD-Gammon methodology of approximation in value space and rollout applies very broadly to deterministic and stochastic optimal control problems. Newton's method here is used for the solution of Bellman's equation, an operator equation that applies universally within DP with both discrete and continuous state and control spaces, as well as finite and infinite horizon. In this connection, we note that the mathematical complications associated with the formalism of Newton's method for nondifferentiable operators have been dealt with in the literature, using sophisticated methods of nonsmooth analysis. We have provided in an appendix a convergence analysis for a finite-dimensional version of Newton's method, which applies to finite-state problems, but conveys clearly the underlying geometrical intuition and points to infinite-state extensions. We have also provided an analysis for the classical linear-quadratic optimal control problem, the associated Riccati equation, and the application of Newton's method for its solution.

While we will deemphasize mathematical proofs in this work, there is considerable related analysis, which supports our conclusions, and can be found in the author's recent RL books [Ber19a], [Ber20a], and the abstract DP monograph [Ber22a]. In particular, the present work may be viewed as a more intuitive, less mathematical, visually oriented exposition of the core material of the research monograph [Ber20a], which deals with approximation in value space, rollout, policy iteration, and multiagent systems. The abstract DP monograph [Ber22a] develops the mathematics that support the visualization framework of the present work, and is a primary resource for followup mathematical research. The RL textbook [Ber19a] provides a more general presentation of RL topics, and includes mathematical proof-based accounts of some of the core material of exact infinite horizon DP, as well as approximate DP, including error bound analyses. Much of this material is also contained, in greater detail, in the author's DP textbook [Ber12]. A mix of material contained in these books forms the core of the author's web-based RL course at ASU.

This monograph, as well as my earlier RL books, were developed while teaching several versions of my course at ASU over the last four years. Videolectures and slides from this course are available from my website

http://web.mit.edu/dimitrib/www/RLbook.html

and provide a good supplement and companion resource to the present book. The hospitable and stimulating environment at ASU contributed much to my productivity during this period, and for this I am very thankful to my colleagues and students for useful interactions. My teaching assistants, Sushmita Bhatacharya, Sahil Badyal, and Jamison Weber, during my courses at ASU have been very supportive. I have also appreciated fruitful discussions with colleagues and students outside ASU, particularly Moritz Diehl, who provided very useful comments on MPC, and Yuchao Li, who proofread carefully the entire book, collaborated with me on research and implementation of various methods, and tested out several algorithmic variants.

Dimitri P. Bertsekas, 2022
dimitrib@mit.edu

1

AlphaZero, Off-Line Training, and On-Line Play

Contents
1.1. Off-Line Training and Policy Iteration . . . . p. 3
1.2. On-Line Play and Approximation in Value Space - Truncated Rollout . . . . p. 6
1.3. The Lessons of AlphaZero . . . . p. 8
1.4. A New Conceptual Framework for Reinforcement Learning . . . . p. 11
1.5. Notes and Sources . . . . p. 14

In this work we will aim to provide a new conceptual framework for reinforcement learning and approximate dynamic programming. These two fields, through the synergy of their ideas in the 1980s and 1990s, and in conjunction of the emergence of machine learning, gave rise to a far-reaching synthesis that would eventually have a major impact on the field of algorithmic optimization.

In this chapter we provide an outline of the motivation and the algorithmic justification of our framework, and its connection to AlphaZero and related game programs, as well as Newton's method for solving fixed point problems. In subsequent chapters, we will flesh out our framework, drawing on the theory of abstract DP, related visualizations, ideas of adaptive, model predictive, and linear quadratic control, as well as paradigms from discrete and combinatorial optimization.

The development of the AlphaZero program by DeepMind Inc, as described in the papers [SHS17], [SSS17], is perhaps the most impressive success story in reinforcement learning (RL) todate. AlphaZero plays Chess, Go, and other games, and is an improvement in terms of performance and generality over the earlier AlphaGo program [SHM16], which plays the game of Go only. AlphaZero, and other chess programs based on similar principles, play as well or better than all competitor computer programs available in 2021, and much better than all humans. These programs are remarkable in several other ways. In particular, they have learned how to play without human instruction, just data generated by playing against themselves. Moreover, they learned how to play very quickly. In fact, AlphaZero learned how to play chess better than all humans and computer programs within hours (with the help of awesome parallel computation power, it must be said).

We should note also that the principles of the AlphaZero design have much in common with the TD-Gammon programs of Tesauro [Tes94], [Tes95], [TeG96] that play backgammon (a game of substantial computational and strategical complexity, which involves a number of states estimated to be in excess of 10^{20}). Tesauro's programs stimulated much interest in RL in the middle 1990s, and one of these programs exhibits similarly different and better play than human backgammon players. A related program for the (one-player) game of Tetris, based on similar principles, is described by Scherrer et al. [SGG15], together with several antecedents, including algorithmic schemes dating to the 1990s, by Tsitsiklis and Van-Roy [TsV96], and Bertsekas and Ioffe [BeI96]. The backgammon and Tetris programs, while dealing with less complex games than chess, are of special interest because they involve significant stochastic uncertainty, and are thus unsuitable for the use of long lookahead minimization, which is widely believed to be one of the major contributors to the success of AlphaZero, and chess programs in general.

Still, for all of their brilliant implementations, these impressive game programs are couched on well established methodology, from optimal and

Sec. 1.1   Off-Line Training and Policy Iteration   3

suboptimal control, which is portable to far broader domains of engineering, economics, and other fields. This is the methodology of dynamic programming (DP), policy iteration, limited lookahead minimization, rollout, and related approximations in value space. The aim of this work is to propose a conceptual, somewhat abstract framework, which allows insight into the connections of AlphaZero and TD-Gammon with some of the core problems in decision and control, and suggests potentially far-reaching extensions.

To understand the overall structure of AlphaZero and related programs, and their connections to the DP/RL methodology, it is useful to divide their design into two parts:

(a) *Off-line training*, which is an algorithm that learns how to evaluate chess positions, and how to steer itself towards good positions with a default/base chess player.

(b) *On-line play*, which is an algorithm that generates good moves in real time against a human or computer opponent, using the training it went through off-line.

An important empirical fact is that *the on-line player of AlphaZero plays much better than its extensively trained off-line player*. This supports a conceptual idea that applies in great generality and is central in this book, namely that *the performance of an off-line trained policy can be greatly improved by on-line play*. We will next briefly describe the off-training and on-line play algorithms, and relate them to DP concepts and principles, focusing on AlphaZero for the most part.

1.1 OFF-LINE TRAINING AND POLICY ITERATION

An off-line training algorithm like the one used in AlphaZero is the part of the program that learns how to play through self-training that takes place before real-time play against any opponent. It is illustrated in Fig. 1.1.1, and it generates a sequence of *chess players* and *position evaluators*. A chess player assigns "probabilities" to all possible moves at any given chess position: these may be viewed as a measure of "effectiveness" of the corresponding moves. A position evaluator assigns a numerical score to any given chess position, and thus predicts quantitatively the performance of a player starting from any position. The chess player and the position evaluator are represented by neural networks, a *policy network* and a *value network*, which accept as input a chess position and generate a set of move probabilities and a position evaluation, respectively.†

---

† Here the neural networks play the role of *function approximators*. By viewing a player as a function that assigns move probabilities to a position, and a position evaluator as a function that assigns a numerical score to a position, the policy and value networks provide approximations to these functions based

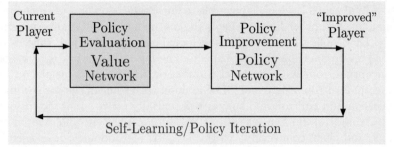

Figure 1.1.1 Illustration of the AlphaZero off-line training algorithm. It generates a sequence of position evaluators and chess players. The position evaluator and the chess player are represented by two neural networks, a value network and a policy network, which accept a chess position and generate a position evaluation and a set of move probabilities, respectively.

In the more conventional DP-oriented terms of this work, a position is the state of the game, a position evaluator is a cost function that gives (an estimate of) the optimal cost-to-go at a given state, and the chess player is a randomized policy for selecting actions/controls at a given state.†

The overall training algorithm is a form of *policy iteration*, a DP algorithm that will be of primary interest to us in this work. Starting from a given player, it repeatedly generates (approximately) improved players, and settles on a final player that is judged empirically to be "best" out of all the players generated. Policy iteration may be separated conceptually into two stages (see Fig. 1.1.1).

(a) *Policy evaluation*: Given the current player and a chess position, the outcome of a game played out from the position provides a single data point. Many data points are thus collected, and are used to train a value network, whose output serves as the position evaluator for that player.

---

on training with data. Actually, AlphaZero uses the same neural network for training both value and policy. Thus there are two outputs of the neural net: value and policy. This is pretty much equivalent to having two separate neural nets and for the purposes of this work, we prefer to explain the structure as two separate networks. AlphaGo uses two separate value and policy networks. Tesauro's backgammon programs use a single value network, and generate moves when needed by one-step or two-step lookahead minimization, using the value network as terminal position evaluator.

† One more complication is that chess and Go are two-player games, while most of our development will involve single-player optimization. While DP theory and algorithms extend to two-player games, we will not discuss these extensions, except in a very limited way in Chapter 6. Alternatively, a chess program can be trained to play well against a fixed opponent, in which case the framework of single-player optimization applies.

Sec. 1.1   Off-Line Training and Policy Iteration

(b) *Policy improvement*: Given the current player and its position evaluator, trial move sequences are selected and evaluated for the remainder of the game starting from many positions. An improved player is then generated by adjusting the move probabilities of the current player towards the trial moves that have yielded the best results.

In AlphaZero (as well as AlphaGo Zero, the version that plays the game of Go) the policy evaluation is done by using deep neural networks. The policy improvement uses a complicated algorithm called *Monte Carlo Tree Search* (MCTS for short), a form of randomized multistep lookahead minimization that enhances the efficiency of the multistep lookahead operation, by pruning intelligently the multistep lookahead graph.

We note, however, that deep neural networks and MCTS, while leading to some performance gains, are not of fundamental importance. The approximation quality that a deep neural network can achieve can also be achieved with a shallow neural network, perhaps with reduced sample efficiency. Similarly MCTS cannot achieve better lookahead accuracy than standard exhaustive search, although it may be more efficient computationally. Indeed, policy improvement can be done more simply without MCTS, as in Tesauro's TD-Gammon program: we try all possible move sequences from a given position, extend forward to some number of moves, and then evaluate the terminal position with the current player's position evaluator. The move evaluations obtained in this way are used to nudge the move probabilities of the current player towards more successful moves, thereby obtaining data that is used to train a policy network that represents the new player.†

Regardless of the use of deep neural networks and MCTS, it is important to note that *the final policy and the corresponding policy evaluation produced by approximate policy iteration and neural network training in AlphaZero involve serious inaccuracies, due to the approximations that are inherent in neural network representations*. The AlphaZero on-line player to be discussed next uses approximation in value space with multistep lookahead minimization, and does not involve any neural network, other than the one that has been trained off-line, so it is not subject to such inaccuracies. As a result, it plays much better than the off-line player.

---

† Quoting from the paper [SSS17] (p. 360): "The AlphaGo Zero selfplay algorithm can similarly be understood as an approximate policy iteration scheme in which MCTS is used for both policy improvement and policy evaluation. Policy improvement starts with a neural network policy, executes a MCTS based on that policy's recommendations, and then projects the (much stronger) search policy back into the function space of the neural network. Policy evaluation is applied to the (much stronger) search policy: the outcomes of selfplay games are also projected back into the function space of the neural network. These projection steps are achieved by training the neural network parameters to match the search probabilities and selfplay game outcome respectively."

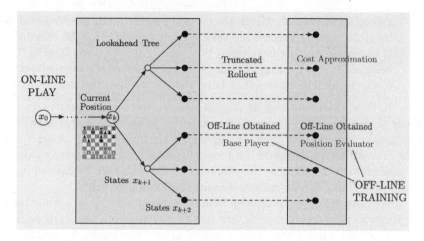

Figure 1.2.1 Illustration of an on-line player such as the one used in AlphaGo, AlphaZero, and Tesauro's backgammon program [TeG96]. At a given position, it generates a lookahead graph of multiple moves up to some depth, then runs the off-line obtained player for some more moves, and evaluates the effect of the remaining moves by using the position evaluator of the off-line player.

1.2 ON-LINE PLAY AND APPROXIMATION IN VALUE SPACE - TRUNCATED ROLLOUT

Consider the "final" player obtained through the AlphaZero off-line training process. It can play against any opponent by generating move probabilities at any position using its off-line trained policy network, and then simply play the move of highest probability. This player would play very fast on-line, but it would not play good enough chess to beat strong human opponents. The extraordinary strength of AlphaZero is attained only after the player obtained from off-line training is embedded into another algorithm, which we refer to as the "on-line player."† In other words *AlphaZero plays on-line much better than the best player it has produced with sophisticated off-line training*. This phenomenon, *policy improvement through on-line play*, is centrally important for our purposes in this work.

Given the policy network/player obtained off-line and its value network/position evaluator, the on-line algorithm plays roughly as follows (see Fig. 1.2.1). At a given position, it generates a lookahead graph of all possi-

---

† Quoting from the paper [SSS17] (p. 354): "The MCTS search outputs probabilities of playing each move. These search probabilities usually select much stronger moves than the raw move probabilities of the neural network." To elaborate, this statement refers to the MCTS algorithm that is used on line to generate the move probabilities at each position encountered in the course of a given game. The neural network referred to is trained off-line, also using in part the MCTS algorithm.

Sec. 1.2   On-Line Play and Approximation in Value Space   7

ble multiple move and countermove sequences, up to a given depth. It then runs the off-line obtained player for some more moves, and evaluates the effect of the remaining moves by using the position evaluator of the value network.

The middle portion, called "truncated rollout," may be viewed as *an economical substitute for longer lookahead minimization*. Actually truncated rollout is not used in the published version of AlphaZero [SHS17]; the first portion (multistep lookahead minimization) is very long and implemented efficiently (partly through the use of MCTS), so that the rollout portion is not essential. Rollout has been used in AlphaGo, the AlphaZero predecessor [SHM16]. Moreover, chess and Go programs (including AlphaZero) typically use a well-known limited form of rollout, called "quiescence search," which aims to resolve imminent threats and highly dynamic positions through simulated multi-move piece exchanges, before invoking the position evaluator. Rollout is instrumental in achieving high performance in Tesauro's 1996 backgammon program [TeG96]. The reason is that backgammon involves stochastic uncertainty, so long lookahead minimization is not possible because of rapid expansion of the lookahead graph with every move.†

In control system design, similar architectures to the ones of AlphaZero and TD-Gammon are employed in model predictive control (MPC). There, the number of steps in lookahead minimization is called the *control interval*, while the total number of steps in lookahead minimization and truncated rollout is called the *prediction interval*; see e.g., Magni et al. [MDM01]. (The MATLAB toolbox for MPC design explicitly allows the user to choose these two intervals.) The benefit of truncated rollout in providing an economical substitute for longer lookahead minimization is well known within this context. We will discuss further the structure of MPC and its similarities with the AlphaZero architecture in Chapter 5.

Dynamic programming frameworks with cost function approximations that are similar to the on-line player illustrated in Fig. 1.2.1, are also known as *approximate dynamic programming*, or *neuro-dynamic pro-*

---

† Tesauro's rollout-based backgammon program [TeG96] uses only a value network, which was trained using an approximate policy iteration scheme developed several years earlier [Tes94]. This network is used to generate moves for the truncated rollout via a one-step or two-step lookahead minimization. Thus the value network also serves as a substitute for the policy network during the rollout operation. The position evaluation used at the end of the truncated rollout is also provided by the value network. The middle portion of Tesauro's scheme (truncated rollout) is important for achieving a very high quality of play, as it effectively extends the length of lookahead from the current position (the player with rollout [TeG96] plays much better than the player without rollout [Tes94]). In backgammon circles, Tesauro's program with truncated rollout is viewed as essentially "optimal."

gramming, and will be central for our purposes. They will be generically referred to as *approximation in value space* in this work.†

Note also that in general, off-line training and on-line policy implementation may be designed independently of each other. For example the off-line training portion may be very simple, such as using a known heuristic policy for rollout without truncation, or without terminal cost approximation. Conversely, a sophisticated process may be used for off-line training of a terminal cost function approximation, which is used following the lookahead minimization in a value space approximation scheme.

1.3 THE LESSONS OF ALPHAZERO

The AlphaZero and TD-Gammon experiences reinforce an important conclusion that applies more generally to decision and control problems: despite the extensive off-line effort that may have gone into the design of a policy, performance may be greatly improved by on-line approximation in value space, with extra lookahead involving minimization and/or with rollout using this policy, and terminal cost approximation.

In the following chapters, we will aim to amplify on this theme and to focus on the principal characteristics of AlphaZero-like architectures, within a broader context of optimal decision and control. We will make use of intuitive visualization, and the central role of Newton's method for solving Bellman's equation.‡ Briefly, our central point will be that *on-line approximation in value space amounts to a step of Newton's method for solving Bellman's equation, while the starting point for the Newton step is based on the results of off-line training*; see Fig. 1.3.1. Moreover, *this starting point may be enhanced by several types of on-line operations, including longer lookahead minimization, and on-line rollout with a policy obtained through off-line training, or heuristic approximations.*

---

† The names "approximate dynamic programming" and "neuro-dynamic programming" are often used as synonyms to RL. However, RL is often thought to also subsume the methodology of approximation in policy space, which involves search for optimal parameters within a parametrized set of policies. The search is done with methods that are largely unrelated to DP, such as for example stochastic gradient or random search methods (see the author's RL textbook [Ber19a]). Approximation in policy space may be used off-line to design a policy that can be used for on-line rollout. However, as a methodological subject, approximation in policy space has little connection to the ideas of the present work.

‡ Bellman's equation, the centerpiece of infinite horizon DP theory, is viewed here as a functional equation, whose solution is the cost of operating the system viewed as a function of the system's initial state. We will give examples of Bellman's equation in Chapter 2 for discounted and other problems, and we will also provide in Chapter 3 abstract forms of Bellman's equation that apply more generally.

Sec. 1.3   The Lessons of AlphaZero

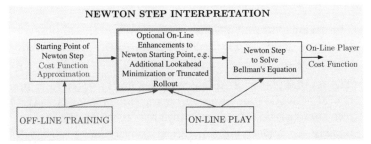

Figure 1.3.1 Illustration of the connections between off-line training, on-line play, and Newton's method for solving Bellman's equation. On-line play is viewed as a Newton step, while off-line training provides the starting point for the Newton step. The Newton step starts with a cost approximation $\tilde{J}$, which may be enhanced on-line by additional lookahead minimization and/or rollout, and produces the cost function of the on-line player.

This interpretation will be the basis for powerful insights into issues of stability, performance, and robustness of the on-line generated policy. In particular, we will aim to show that feedback control, based on approximation in value space and the underlying off-line training/on-line play structure, offers benefits that go well beyond the conventional wisdom that "feedback corrects for uncertainty, and modeling errors." The reason is that by overlaying on-line play on top of off-line training, we gain significantly in performance, by correcting (through the Newton step) for the errors that are inherent in off-line training with approximation architectures such as neural networks.

Our mathematical framework is couched on unifying principles of abstract DP, including abstract forms of Bellman's equation, and the value and policy iteration algorithms (see the author's books [Ber12], [Ber22a]). However, in this work, we will deemphasize mathematical proofs. There is considerable related analysis, which supports our conclusions and can be found in the author's recent RL books [Ber19a], [Ber20a].

In summary, our discussion will aim to highlight the following points:

Summary

(a) Approximation in value space, with one-step lookahead minimization, is an exact step of Newton's method for solving Bellman's equation. This step may be preceded by on-line adjustments and/or value iterations, which enhance its starting point.

(b) The starting point for the Newton step of (a) is obtained by some unspecified off-line methodology, which may involve the solution of a related but simpler problem, and/or training with data that makes use of neural networks or feature-based architectures.

(c) The on-line play and off-line training parts of the AlphaZero/TD-Gammon design structure correspond to (a) and (b) above, respectively.

(d) The on-line player of AphaZero plays much better than its deep neural network-trained player for the same reason that the Newton step (a) improves substantially on its starting point (b), namely the underlying superlinear convergence property that is typical of Newton's method.

(e) ℓ-step lookahead minimization can be viewed as one-step lookahead minimization where $\ell - 1$ value iterations are used to enhance the starting point of the Newton step of (a) above. It is important to perform the first step of the lookahead exactly, but for the last $\ell - 1$ steps, approximations may be tolerated.

(f) The algorithmic processes for (a) and (b) above can be designed by a variety of methods, and independently of each other. For example:

  (1) The implementation of the Newton step (a) may or may not involve any of the following: truncated rollout, on-line Monte Carlo simulation, MCTS or other efficient graph search techniques, forms of continuous space optimization, on-line policy iteration, etc.

  (2) The computation of the starting point (b) may or may not involve any of the following: Q-learning, approximate policy iteration based on temporal differences or aggregation, neural networks, feature-based function approximation, policies trained off-line by approximation in policy space, including policy gradient methods or policy random search, etc. Moreover, the details of this computation may vary broadly without affecting significantly the effectiveness of the overall scheme, which is primarily determined by the Newton step (a).

(g) An efficient implementation of the Newton step (a) is often critical in order to meet real-time constraints for generating controls, and to allow longer lookahead minimization, which enhances the starting point of the Newton step and its performance. By contrast, off-line training algorithms used for (b) have much less stringent real-time constraints, and the issues of sample efficiency and fine tuned performance, while important, are not critical.

(h) The efficient implementation of the Newton step may benefit from the use of distributed computation and other simplifications. A case in point is multiagent problems, which we will discuss later (see Chapter 3).

(i) Approximation in value space addresses effectively issues of robustness and on-line replanning for problems with changing parameters. The mechanism is similar to the one of indirect adaptive control: changing problem parameters are estimated on-line and a Newton step is used in place of an expensive full reoptimization of the controller. In the presence of changing parameters, the Bellman equation changes, but the Newton step itself remains powerful and aims at the optimal solution that corresponds to the estimated system parameters.

(j) Model predictive control (MPC) has a conceptually similar structure to the AlphaZero-like programs, and entails an on-line play component involving multistep lookahead minimization, forms of truncated rollout, and an off-line training component to construct terminal cost approximations, and "safe" state space regions or reachability tubes to deal with state constraints. The success of MPC may be attributed to these similarities and to its resilience to changing problem parameters as per (i) above.

(k) On-line rollout with a stable policy yields a good starting point for the Newton step (a): it improves the stability properties of the policy obtained by approximation in value space, and provides an economical substitute for long lookahead minimization.

(l) Because the ideas outlined above are couched on principles of DP that often hold for arbitrary state and control spaces, they are valid within very general contexts: continuous-spaces control systems, discrete-spaces Markov decision problems, control of hybrid systems, decision making in multiagent systems, and discrete and combinatorial optimization.

The preceding points are meant to highlight the essence of the connections between AlphaZero and TD-Gammon, approximation in value space, and decision and control. Naturally in practice there are exceptions and modifications, which need to be worked out in the context of particular applications, under appropriate assumptions. Moreover, while some results and elaborations are available through the research that has been done on approximate DP and on MPC, several of the results suggested by the analysis and insights of the present work remain to be rigorously established and enhanced within the context of specific problems.

1.4 A NEW CONCEPTUAL FRAMEWORK FOR REINFORCEMENT LEARNING

In this work we will emphasize the distinct roles of off-line training and on-line play algorithms within the structure of approximate sequential decision making and approximation in value space schemes. In doing so, we will aim for a new conceptual framework for RL, which is based on the synergism and complementarity of off-line training and on-line play, and the analytical framework of Newton's method.

We will implicitly assume that *the time available for off-line training is very long (practically limitless)*, but that the problem at hand is such that exact DP algorithms, like policy iteration and Q-learning, are impossible for one (or both) of the following two reasons:

(a) There are too many states (either an infinite number as in continuous space problems, or very large as in chess). As a result a lookup table representation of policies, value functions, and/or Q-factors is impossible, and the only practical alternative is a compact representation, via a neural network or some other approximation architecture.

(b) The system model is changing over time as in adaptive control, so even if an exactly optimal policy is computed off-line under some nominal problem assumptions, it becomes suboptimal when the problem parameters change.

In this work, we will not discuss training algorithms and their associated sample efficiency issues, and we will refer to the many available sources, including the author's RL books [Ber19a], [Ber20a].

On the other hand, we will assume that there is limited time for on-line decision making, because of hard practical constraints on the real time that is available between decisions. These constraints are highly problem dependent: for some problems, following the observation of the state, we may need to produce the next decision within a fraction of a second, whereas for others we may have hours at our disposal. We will assume that *whatever time is available, it will be used to provide quite accurate (nearly exact) one-step or multistep lookahead minimization, and time permitting, to extend as much as possible the combined length of the lookahead minimization and the truncated rollout with an off-line computed policy.* We will thus implicitly take it as given that longer (as well as more accurate) lookahead minimization is better for the performance of the policy obtained,† although the division of effort between lookahead minimization and truncated rollout with a policy is a design decision that may depend on the circumstances. Note that parallel and distributed computation can play an important role in mitigating practical on-line time constraints.

The central fact in our conceptual framework is that approximation in value space with one-step lookahead minimization constitutes a single Newton step for solving Bellman's equation. Contrary to other Newton-like steps that may have been part of the off-line training process, this

---

† It is possible to construct artificial problems, where longer lookahead results in worse performance (see [Ber19a], Section 2.2), but such problems are rare in practice.

single Newton step is accurate: *all the approximation has been shifted to its starting point*. Moreover, the Newton step can be very powerful, and its starting point can be enhanced by multistep lookahead minimization or by truncated rollout. From an algorithmic point of view, the Newton step converges superlinearly without the need for differentiability of the Bellman operator T: it takes advantage of the monotonicity and concavity structure of T (see the Appendix, where we will discuss Newton's method without differentiability assumptions).

To summarize, both off-line training and on-line play are subject to fundamental limits: the former's limit is the constrained power of the approximation architecture, while the latter's limit is the constrained on-line computation time. The former limit cannot be easily overcome, but the latter limit can be stretched a lot thanks to the power of the Newton step, supplemented by long lookahead minimization and truncated rollout, as well as through the use of parallel and distributed computation.

Our design philosophy in a nutshell is the following:

(1) The major determinant of the quality of the controller obtained by our schemes is the Newton step that is performed on-line. Stated somewhat bluntly, *off-line training is secondary by comparison*, in the sense that without on-line one-step or multistep lookahead minimization, the quality of the policy obtained by off-line training alone is often unacceptably poor. In particular, whether done by neural networks, feature-based linear architectures, temporal difference methods, aggregation, policy gradient, policy random search, or whatever other reasonable approach, off-line training principally serves the purpose of providing a good or reasonable starting point for the Newton step. This is the principal lesson from AlphaZero and TD-Gammon in our view. This philosophy also underlies MPC, where on-line lookahead minimization has traditionally been the principal focus, perhaps supplemented by truncated rollout, with off-line calculations playing a limited subsidiary role.†

(2) The Newton step is often powerful enough to smooth out differences in various off-line training methods. In particular, methods such as TD(λ) with different values of λ, policy gradient, linear programming, etc, all give different, but more or less equally good starting points for the Newton step. The conclusion from this is that off-line training with a very large number of samples, and sophisticated ways to improve sample efficiency may not be very useful, beyond a certain point, because gains in efficiency and accuracy tend to be washed up by the Newton step.

---

† Incidentally, this is a major reason why there is an apparent disconnect between the MPC community, which is mostly focused on on-line play, and the RL community, which is mostly focused on off-line training.

(3) The on-line Newton step also works well in the context of adaptive control, as long as it is calculated on the basis of the currently correct model parameters (so this requires an on-line parameter identification algorithm). The reason is that when problem parameters change, the Bellman operator changes, but the Newton step is executed on the basis of the correct Bellman operator. This is also a principal reason why MPC schemes have been used with success in adaptive control contexts.

We will return to these points repeatedly in the course of our presentation.

1.5 NOTES AND SOURCES

The theory of DP dates to the late 40s and 50s, and provides the foundation for our subject. Indeed RL may be viewed as an approximate form of the exact DP methodology. The author's DP textbook [Ber17a] provides an extensive discussion of finite horizon DP, and its applications to discrete and continuous spaces problems, using a notation and style that is consistent with the present book. The books by Puterman [Put94] and by the author [Ber12] provide detailed treatments of infinite horizon finite-state Markovian decision problems.

Continuous spaces infinite horizon problems are covered in the author's book [Ber12], while some of the more complex mathematical aspects of exact DP are discussed in the monograph by Bertsekas and Shreve [BeS78] (particularly the probabilistic/measure-theoretic issues associated with stochastic optimal control).†

---

† The rigorous mathematical theory of stochastic optimal control, including the development of an appropriate measure-theoretic framework, dates to the 60s and 70s. It culminated in the monograph [BeS78], which provides the now "standard" framework, based on the formalism of Borel spaces, lower semianalytic functions, and universally measurable policies. This development involves daunting mathematical complications, which stem, among others, from the fact that when a Borel measurable function $F(x, u)$, of the two variables x and u, is minimized with respect to u, the resulting function

$$G(x) = \min_u F(x, u)$$

need not be Borel measurable (it is lower semianalytic). Moreover, even if the minimum is attained by several functions/policies μ, i.e., $G(x) = F\bigl(x, \mu(x)\bigr)$ for all x, it is possible that none of these μ is Borel measurable (however, there does exist a minimizing policy that belongs to the broader class of universally measurable policies). Thus, starting with a Borel measurability framework for cost functions and policies, we quickly get outside that framework when executing DP algorithms, such as value and policy iteration. The broader framework of universal measurability is required to correct this deficiency, in the absence of additional (fairly strong) assumptions.

Sec. 1.5   Notes and Sources   15

The third edition of the author's abstract DP monograph [Ber22a], expands on the original 2013 first edition, and aims at a unified development of the core theory and algorithms of total cost sequential decision problems. It addresses simultaneously stochastic, minimax, game, risk-sensitive, and other DP problems, through the use of abstract DP operators (or Bellman operators as they are often called in RL). The abstract framework is important for some the visualization insights and the connections to Newton's method that are central for the purposes of this book.

The approximate DP and RL literature has expanded tremendously since the connections between DP and RL became apparent in the late 1980s and early 1990s. In what follows, we will provide a list of textbooks, research monographs, and broad surveys, which supplement our discussions, express related viewpoints, and collectively provide a guide to the literature.

RL Textbooks

Two books were written in the 1990s, setting the tone for subsequent developments in the field. One in 1996 by Bertsekas and Tsitsiklis [BeT96], which reflects a decision, control, and optimization viewpoint, and another in 1998 by Sutton and Barto, which reflects an artificial intelligence viewpoint (a 2nd edition, [SuB18], was published in 2018). We refer to the former book and also to the author's DP textbooks [Ber12], [Ber17a] for a broader discussion of some of the topics of this book, including algorithmic convergence issues and additional DP models, such as those based on average cost and semi-Markov problem optimization. Note that both of these books deal with finite-state Markov decision models and use a transition probability notation, as they do not address continuous spaces problems, which are also of major interest in this book.

More recent books are by Gosavi [Gos15] (a much expanded 2nd edition of his 2003 monograph), which emphasizes simulation-based optimization and RL algorithms, Cao [Cao07], which focuses on a sensitivity approach to simulation-based methods, Chang, Fu, Hu, and Mar-

---

The monograph [BeS78] provides an extensive treatment of these issues, while Appendix A of the DP textbook [Ber12] provides a tutorial introduction. The followup work by Huizhen Yu and the author [YuB15] resolves the special measurability issues that relate to policy iteration, and provides additional analysis relating to value iteration. In the RL literature, the mathematical difficulties around measurability are usually neglected (as they are in the present book), and this is fine because they do not play an important role in applications. Moreover, measurability issues do not arise for problems involving finite or countably infinite state and control spaces. We note, however, that there are quite a few published works in RL as well as exact DP, which purport to address measurability issues with a mathematical narrative that is either confusing or plain incorrect.

cus [CFH13] (a 2nd edition of their 2007 monograph), which emphasizes finite-horizon/multistep lookahead schemes and adaptive sampling, Busoniu, Babuska, De Schutter, and Ernst [BBD10a], which focuses on function approximation methods for continuous space systems and includes a discussion of random search methods, Szepesvari [Sze10], which is a short monograph that selectively treats some of the major RL algorithms such as temporal difference methods, armed bandit methods, and Q-learning, Powell [Pow11], which emphasizes resource allocation and operations research applications, Powell and Ryzhov [PoR12], which focuses on specialized topics in learning and Bayesian optimization, Vrabie, Vamvoudakis, and Lewis [VVL13], which discusses neural network-based methods and on-line adaptive control, Kochenderfer et al. [KAC15], which selectively discusses applications and approximations in DP, and the treatment of uncertainty, Jiang and Jiang [JiJ17], which addresses adaptive control and robustness issues within an approximate DP framework, Liu, Wei, Wang, Yang, and Li [LWW17], which deals with forms of adaptive dynamic programming, and topics in both RL and optimal control, and Zoppoli, Sanguineti, Gnecco, and Parisini [ZSG20], which addresses neural network approximations in optimal control as well as multiagent/team problems with nonclassical information patterns.

There are also several books that, while not exclusively focused on DP and/or RL, touch upon several of the topics of this book. The book by Borkar [Bor08] is an advanced monograph that addresses rigorously many of the convergence issues of iterative stochastic algorithms in approximate DP, mainly using the so called ODE approach. The book by Meyn [Mey07] is broader in its coverage, but discusses some of the popular approximate DP/RL algorithms. The book by Haykin [Hay08] discusses approximate DP in the broader context of neural network-related subjects. The book by Krishnamurthy [Kri16] focuses on partial state information problems, with discussion of both exact DP, and approximate DP/RL methods. The textbooks by Kouvaritakis and Cannon [KoC16], Borrelli, Bemporad, and Morari [BBM17], and Rawlings, Mayne, and Diehl [RMD17] collectively provide a comprehensive view of the MPC methodology. The book by Lattimore and Szepesvari [LaS20] is focused on multiarmed bandit methods. The book by Brandimarte [Bra21] is a tutorial introduction to DP/RL that emphasizes operations research applications and includes MATLAB codes. The book by Hardt and Recht [HaR21] focuses on broader subjects of machine learning, but covers selectively approximate DP and RL topics as well.

The present book is similar in style, terminology, and notation to the author's recent RL textbooks [Ber19a], [Ber20a], and the 3rd edition of the abstract DP monograph [Ber22a], which collectively provide a fairly comprehensive account of the subject. In particular, the 2019 RL textbook includes a broader coverage of approximation in value space methods, including certainty equivalent control and aggregation methods. It also covers substantially policy gradient methods for approximation in policy space, which we will not address here. The 2020 book focuses more closely on rollout, policy iteration, and multiagent problems. The abstract DP monograph [Ber22a] is an advanced treatment of exact DP, which also connects with intuitive visualizations of Bellman's equation and related algorithms. The present book is less mathematical and more conceptual in character. It focuses on the connection of approximation in value space with Newton's method, relying on analysis first provided in the book [Ber20a] and the paper [Ber22b], as well as on visualizations of abstract DP ideas from the book [Ber22a].

Surveys and Short Research Monographs

In addition to textbooks, there are many surveys and short research monographs relating to our subject, which are rapidly multiplying in number. Influential early surveys were written, from an artificial intelligence viewpoint, by Barto, Bradtke, and Singh [BBS95] (which dealt with the methodologies of real-time DP and its antecedent, real-time heuristic search [Kor90], and the use of asynchronous DP ideas [Ber82], [Ber83], [BeT89] within their context), and by Kaelbling, Littman, and Moore [KLM96] (which focused on general principles of RL). The volume by White and Sofge [WhS92] also contains several surveys describing early work in the field.

Several overview papers in the volume by Si, Barto, Powell, and Wunsch [SBP04] describe some approximation methods that we will not be covering in much detail in this book: linear programming approaches (De Farias [DeF04]), large-scale resource allocation methods (Powell and Van Roy [PoV04]), and deterministic optimal control approaches (Ferrari and Stengel [FeS04], and Si, Yang, and Liu [SYL04]). Updated accounts of these and other related topics are given in the survey collections by Lewis, Liu, and Lendaris [LLL08], and Lewis and Liu [LeL13].

Recent extended surveys and short monographs are Borkar [Bor09] (a methodological point of view that explores connections with other Monte Carlo schemes), Lewis and Vrabie [LeV09] (a control theory point of view), Szepesvari [Sze10] (which discusses approximation in value space from a RL point of view), Deisenroth, Neumann, and Peters [DNP11], and Grondman et al. [GBL12] (which focus on policy gradient methods), Browne et al. [BPW12] (which focuses on Monte Carlo Tree Search), Mausam and Kolobov [MaK12] (which deals with Markov decision problems from an artificial intelligence viewpoint), Schmidhuber [Sch15], Arulkumaran et al. [ADB17], Li [Li17], Busoniu et al. [BDT18], the author's [Ber05] (which focuses on rollout algorithms and model predictive control), [Ber11] (which focuses on approximate policy iteration), and [Ber18b] (which focuses on aggregation methods), and Recht [Rec18] (which focuses on continuous spaces optimal control).

Research Content of this Book

The research focus of this book is to propose and develop a new conceptual framework, which the author believes is fundamental within the context of DP-based RL methodology. This framework centers around the division of the design process of an RL scheme into the off-line training and the on-line play algorithms, and shows that these algorithms operate in synergy through the powerful mechanism of Newton's method.

The style of the book is different than the style of the author's more mathematically oriented RL books [Ber19a] and [Ber20a], and abstract DP book [Ber22a]. In particular, the present book emphasizes insights through visualization rather than rigorous proofs. At the same time, the book makes careful distinctions between provable and speculative claims. By highlighting the exceptional behavior that may occur, the book also aims to emphasize the need for serious mathematical research and experimentation into broad classes of problems, beyond the relatively well-behaved finite horizon and discounted/contractive problems.

Book Organization

The book is structured as follows. In Chapter 2, we review the theory of classical infinite horizon optimal control problems, in order to provide some orientation and an analytical platform for what follows in subsequent chapters. In Chapter 3, we introduce an abstract DP framework that will set the stage for the conceptual and visual interpretations of approximation in value space in terms of Newton's method. In this chapter, we also present new research relating to on-line policy iteration, which aims to improve the on-line approximation in value space algorithm by using training data that is collected on-line. In Chapter 4, we illustrate our analysis within the simple and intuitive framework of linear quadratic problems, which admit visualizations through the Riccati equation operators. In Chapter 5, we discuss various issues of changing problem parameters, adaptive control, and MPC. In Chapter 6, we extend the ideas of earlier chapters to finite horizon problems and discrete optimization, with a special focus on rollout algorithms and their variations. This chapter also includes a section on approximation in value space schemes for deterministic continuous-time optimal control. Finally, in the Appendix, we outline the convergence theory of Newton's method, and explain how the theory applies to nondifferentiable fixed point problems, such as the solution of Bellman's equation in DP. We also describe how the connection with Newton's method can be used to derive new and more realistic error bounds for approximation in value space and approximate policy iteration.

2

Deterministic and Stochastic Dynamic Programming

Contents
2.1. Optimal Control Over an Infinite Horizon . . . . p. 20
2.2. Approximation in Value Space . . . . p. 25
2.3. Notes and Sources . . . . p. 30

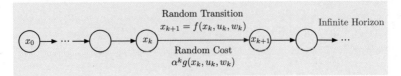

Figure 2.1.1 Illustration of an infinite horizon problem. The system and cost per stage are stationary, except for the use of a discount factor α. If $\alpha = 1$, there is typically a special cost-free termination state that we aim to reach.

In this chapter we will describe a classical framework of optimal control over an infinite horizon. We will use it as a principal example for a more abstract DP framework to be introduced in Chapter 3. This abstract framework will be used in turn as the starting point for our analysis and visualization of algorithmic issues, relating to approximation in value space, multistep lookahead, controller stability, truncated rollout, and policy iteration. Note that finite horizon problems can be converted to the infinite horizon format of this chapter, as will be discussed in Chapter 6.

2.1 OPTIMAL CONTROL OVER AN INFINITE HORIZON

Let us consider a familiar class of stochastic optimal control problems over an infinite horizon (see Fig. 2.1.1). We have a stationary system of the form

$$x_{k+1} = f(x_k, u_k, w_k), \qquad k = 0, 1, \ldots,$$

where x_k is an element of some state space X and the control u_k is an element of some control space U; see Fig. 2.1.1. The system includes a random "disturbance" w_k, taking values in some space W, with a probability distribution $P(\cdot \mid x_k, u_k)$ that may depend explicitly on x_k and u_k, but not on values of prior disturbances $w_{k-1}, \ldots, w_0$.† The control u_k is constrained to take values in a given subset $U(x_k) \subset U$, which depends on the current state x_k. We are interested in policies $\pi = \{\mu_0, \mu_1, \ldots\}$, such that each function μ_k maps states into controls, and satisfies $\mu_k(x_k) \in U(x_k)$ for all k. A stationary policy of the form $\{\mu, \mu, \ldots\}$ will also be referred to as "policy μ." We make no assumptions on the state, control, and disturbances, and indeed for most of the discussion of this work, these spaces can be arbitrary.

We aim to minimize the expected total cost over an infinite number of stages, given by

$$J_\pi(x_0) = \lim_{N \to \infty} E\left\{\sum_{k=0}^{N-1} \alpha^k g(x_k, \mu_k(x_k), w_k)\right\}, \qquad (2.1)$$

---

† We assume an introductory probability background on the part of the reader. For an account that is consistent with our use of probability in this book, see the text by Bertsekas and Tsitsiklis [BeT08].

Sec. 2.1 Optimal Control Over an Infinite Horizon

where $\alpha^k g(x_k, u_k, w_k)$ is the cost of stage k, and $\alpha \in (0, 1]$ is a discount factor. If $\alpha = 1$ we refer to the problem as undiscounted. The expected value in Eq. (2.1) is taken with respect to the random disturbances w_k, $k = 0, 1, \ldots$. Here, $J_\pi(x_0)$ denotes the cost associated with an initial state x_0 and a policy $\pi = \{\mu_0, \mu_1, \ldots\}$. The cost function of a stationary policy μ is denoted by J_μ. The optimal cost starting at state x, $\inf_\pi J_\pi(x)$, is denoted by $J^*(x)$, and the function J^* is referred to as the *optimal cost function*.

Let us consider some special cases, which will be of primary interest in this work:

(a) *Stochastic shortest path problems* (SSP for short). Here, $\alpha = 1$ but there is a special cost-free termination state, denoted by t; once the system reaches t it remains there at no further cost. Usually, the termination state t represents a goal state that we are trying to reach at minimum cost; these are problems where the cost per stage is nonnegative, and will be of primary interest in this work. In some other types of problems, t may be a state that we are trying to avoid for as long as possible; these are problems where the cost per stage is nonpositive, and will not be specifically discussed in this work.

(b) *Discounted stochastic problems.* Here, $\alpha < 1$ and there need not be a termination state. However, there is a substantial connection between SSP and discounted problems. Aside from the fact that they are both infinite horizon total cost optimization problems, a discounted problem can be readily converted to an SSP problem. This can be done by introducing an artificial termination state to which the system moves with probability $1 - \alpha$ at every state and stage, thus making termination inevitable. Thus SSP and discounted problems share qualitative similarities in their respective theories.

(c) *Deterministic nonnegative cost problems.* Here, the disturbance w_k takes a single known value. Equivalently, there is no disturbance in the system equation and the cost expression, which now take the form

$$x_{k+1} = f(x_k, u_k), \qquad k = 0, 1, \ldots, \tag{2.2}$$

and

$$J_\pi(x_0) = \lim_{N \to \infty} \sum_{k=0}^{N-1} \alpha^k g(x_k, \mu_k(x_k)). \tag{2.3}$$

We assume further that there is a cost-free and absorbing termination state t, and we have

$$g(x, u) \geq 0, \qquad \text{for all } x \neq t, \ u \in U(x), \tag{2.4}$$

and $g(t, u) = 0$ for all $u \in U(t)$. This type of structure expresses the objective to reach or approach t at minimum cost, a classical control problem. An extensive analysis of the undiscounted version of this problem was given in the author's paper [Ber17b].

An important special case is *finite-state deterministic problems*. Finite horizon versions of these problems include challenging discrete optimization problems, whose exact solution is practically impossible. It is possible to transform such problems to infinite horizon SSP problems, so that the conceptual framework developed here applies (see Chapter 6). The approximate solution of discrete optimization problems by RL methods, and particularly by rollout, has been discussed at length in the books [Ber19a] and [Ber20a].

Another important deterministic nonnegative cost problem is the classical continuous spaces problem where the system is linear, with no control constraints, and the cost function is quadratic; see the following example. We will often refer to this problem and its extensions in what follows.

Example 2.1.1 (Linear Quadratic Problems)

Assume that the system is linear of the form

$$x_{k+1} = Ax_k + Bu_k, \qquad (2.5)$$

where x_k and u_k are elements of the Euclidean spaces $\Re^n$ and $\Re^m$, respectively, A is an $n \times n$ matrix, and B is an $n \times m$ matrix. We assume that there are no control constraints. The cost per stage is quadratic of the form

$$g(x, u) = x'Qx + u'Ru, \qquad (2.6)$$

where Q and R are positive definite symmetric matrices of dimensions $n \times n$ and $m \times m$, respectively (all finite-dimensional vectors in this work are viewed as column vectors, and a prime denotes transposition). It is well known that this problem admits a nice analytical solution, which we will discuss shortly, and we will use later for illustrations, examples, and counterexamples (see also [Ber17a], Section 3.1).

Infinite Horizon Methodology

Many of the analytical and computational issues regarding infinite horizon problems revolve around the relation between the optimal cost function J^* of the problem and the optimal cost function of the corresponding N-stage problem. In particular, let $J_N(x)$ denote the optimal cost of the problem involving N stages, initial state x, cost per stage $g(x, u, w)$, and zero terminal cost. This cost is generated at the Nth iteration of the *value iteration* algorithm (VI for short)

$$J_{k+1}(x) = \min_{u \in U(x)} E\Big\{g(x, u, w) + \alpha J_k\big(f(x, u, w)\big)\Big\}, \qquad k = 0, 1, \ldots, \quad (2.7)$$

starting from $J_0(x) \equiv 0$ (see Chapter 6). It is natural to speculate the following three basic properties:†

(1) The optimal infinite horizon cost is the limit of the corresponding N-stage optimal costs as $N \to \infty$:

$$J^*(x) = \lim_{N \to \infty} J_N(x) \qquad (2.8)$$

for all states x.

(2) Bellman's equation holds:

$$J^*(x) = \min_{u \in U(x)} E\Big\{ g(x, u, w) + \alpha J^*\big(f(x, u, w)\big) \Big\}, \qquad \text{for all } x. \quad (2.9)$$

This equation can be viewed as the limit as $k \to \infty$ of the VI algorithm (2.7), assuming property (1) above holds and guarantees that $J_k(x) \to J^*(x)$ for all x. There is also a Bellman equation for each stationary policy μ. It is given by

$$J_\mu(x) = E\Big\{ g(x, \mu(x), w) + \alpha J_\mu\big(f(x, \mu(x), w)\big) \Big\}, \qquad \text{for all } x, \quad (2.10)$$

where J_μ is the cost function of μ. We can view this as just the Bellman equation (2.9) for a different problem, where for each x, the control constraint set $U(x)$ consists of just one control, namely $\mu(x)$.

(3) If $\mu(x)$ attains the minimum in the right-hand side of the Bellman equation (2.9) for each x, then the stationary policy μ should be optimal.

All three of the preceding results hold for discounted problems, provided the expected cost per stage $E\{g(x, u, w)\}$ is bounded over the set of possible values of (x, u, w) (see the DP book [Ber12], Chapter 1). They also hold for finite-state SSP problems under reasonable assumptions. For deterministic problems with possibly infinite state and control spaces, there is substantial analysis that provides assumptions under which the results (1)-(3) above hold (see e.g., [Ber12]).

The VI algorithm is also typically valid, in the sense that $J_k \to J^*$, even if the initial function J_0 is nonzero. The motivation for a different choice of J_0 is faster convergence to J^*; generally the convergence is faster as J_0 is chosen closer to J^*. The intuitive interpretation of the Bellman equation (2.9) is that it is the limit as $k \to \infty$ of the VI algorithm (2.7) assuming that $J_k \to J^*$. The optimality condition (3) indicates that optimal and near optimal policies can be obtained from within the class of

---

† Throughout this work, we will be using "min" instead of the more formal "inf," even if we are not sure that the minimum is attained.

stationary policies, something that is generally true for the problems that we discuss in this work, and that we will implicitly assume in what follows.

Aside from the VI algorithm, another fundamental algorithm is policy iteration (PI for short), which will be discussed in Section 3.3. It is much faster than VI, and in practice it often requires a handful of iterations, independently of the problem size. An explanation is that PI can be viewed as Newton's method for solving Bellman's equation, as we will explain in Section 3.3. This connection with Newton's method extends to approximate forms of PI, and is central for our purposes in this book.

The author's paper [Ber17b], and also the abstract DP book [Ber22a], provide a detailed analysis of the undiscounted special case of the problem (2.2)-(2.4), where there is a cost-free and absorbing termination state t, the cost function is strictly positive for all other states, as in Eq. (2.4), and the objective is to reach or asymptotically approach the termination state. This analysis covers the preceding four properties, as well as the issue of convergence of PI, for the case of general state and control spaces (continuous or discrete or a mixture thereof). It delineates conditions under which favorable properties can be guaranteed.

Example 2.1.1 (Linear Quadratic Problems - Continued)

Consider again the linear quadratic problem defined by Eqs. (2.5)-(2.6). The Bellman equation is given by

$$J(x) = \min_{u \in \Re^m} \{x'Qx + u'Ru + J(Ax + Bu)\}, \tag{2.11}$$

and turns out to have a unique solution within the space of quadratic functions of the form

$$J(x) = x'Kx, \tag{2.12}$$

where K is a positive semidefinite symmetric matrix [under our positive definiteness assumption on Q and R, and an additional controllability assumption on the system (2.5); see [Ber17a], Section 3.1]. This unique solution can be shown to be the optimal cost function of the problem, and has the form

$$J^*(x) = x'K^*x. \tag{2.13}$$

We can obtain K^* by solving the matrix equation

$$K = F(K), \tag{2.14}$$

with $F(K)$ defined over symmetric matrices K by

$$F(K) = A'\big(K - KB(B'KB + R)^{-1}B'K\big)A + Q. \tag{2.15}$$

The optimal policy is obtained by minimization in the Bellman equation (2.11) when J is replaced by the optimal cost function J^* of Eq. (2.13). It can be verified that it is linear of the form

$$\mu^*(x) = Lx,$$

where L is the matrix

$$L = -(B'K^*B + R)^{-1} B'K^* A.$$

The VI and PI algorithms are known to have favorable properties for our linear quadratic problem. In particular, the VI algorithm can be executed within the space of positive semidefinite symmetric matrices. The VI algorithm $J_{k+1} = TJ_k$, when J_k has the form $J_k(x) = x'K_k x$, yields for all x,

$$J_{k+1}(x) = x'K_{k+1}x \quad \text{with} \quad K_{k+1} = F(K_k), \tag{2.16}$$

where F is given by Eq. (2.15). It can be shown that the sequence $\{K_k\}$ converges to the optimal matrix K^* starting from any positive semidefinite symmetric matrix K_0 under the assumptions mentioned earlier. The PI algorithm also has favorable convergence properties (under the same assumptions) and its important connection with Newton's method will be discussed later.

The preceding results are well known and they are given with proofs in several control theory texts, including the author's DP books [Ber17a], Chapter 3, and [Ber12], Chapter 4.† The equation $K = F(K)$ is known as the *Riccati equation*.‡ It can be viewed as the Bellman equation restricted to the subspace of quadratic functions of the form (2.12). Note that the Riccati equation can be shown to have solutions other than K^* (which necessarily are not positive definite symmetric). Illustrative examples will be given later.

2.2 APPROXIMATION IN VALUE SPACE

A principal RL approach to deal with the often intractable exact computation of J^* is *approximation in value space*. Here in place of J^*, we use an approximation $\tilde{J}$, and generate at any state x, a control $\tilde{\mu}(x)$ by the *one-step lookahead minimization*

$$\tilde{\mu}(x) \in \arg\min_{u \in U(x)} E\Big\{g(x, u, w) + \alpha \tilde{J}\big(f(x, u, w)\big)\Big\}; \tag{2.17}$$

---

† Actually the preceding formulas also hold even when the positive definiteness assumption on Q is replaced by other weaker conditions (see [Ber17a], Section 3.1). We will not go into the details of this, but we note that some condition on Q is needed for the preceding results to hold, as we will show later by example in Chapter 4.

‡ This is an algebraic form of the Riccati differential equation, which was invented in its one-dimensional form by count Jacopo Riccati in the 1700s, and has played an important role in control theory. It has been studied extensively in its differential and difference matrix versions; see the book by Lancaster and Rodman [LaR95], and the paper collection by Bittanti, Laub, and Willems [BLW91], which also includes a historical account by Bittanti [Bit91] of Riccati's remarkable life and accomplishments.

(we implicitly assume that the minimum above is attained for all x).† This minimization yields a stationary policy $\{\tilde{\mu}, \tilde{\mu}, \ldots\}$, with cost function denoted $J_{\tilde{\mu}}$ [i.e., $J_{\tilde{\mu}}(x)$ is the total infinite horizon discounted cost obtained when using $\tilde{\mu}$ starting at state x]. In the next section, *the change from $\tilde{J}$ to $J_{\tilde{\mu}}$ will be interpreted as a step of Newton's method for solving Bellman's equation*. Among others, this will suggest that $J_{\tilde{\mu}}$ is close to J^* and obeys a superlinear convergence relation

$$\lim_{\tilde{J} \to J^*} \frac{J_{\tilde{\mu}}(x) - J^*(x)}{\tilde{J}(x) - J^*(x)} = 0,$$

for all states x. For specific types of problems, this relation represents a plausible result, which likely holds under appropriate conditions. This is similar to the use of Newton's method in numerical analysis, where its global or local convergence is guaranteed only under some assumptions. Within our context of approximate DP, however, *there is an important underlying structure, which is favorable and enhances the convergence properties of Newton's method, namely the monotonicity and concavity properties of Bellman's equation*, as we will discuss in what follows.

While it is desirable that $J_{\tilde{\mu}}$ is close to J^* in some sense, for classical control problems involving control to a goal state (e.g., problems with a cost-free and absorbing terminal state, and positive cost for all other states), stability of $\tilde{\mu}$ may be a principal objective. For the purposes of this work, we will focus on stability issues for just this one class of problems, and *we will consider the policy $\tilde{\mu}$ to be stable if $J_{\tilde{\mu}}$ is real-valued*, i.e.,

$$J_{\tilde{\mu}}(x) < \infty, \qquad \text{for all } x \in X.$$

Selecting $\tilde{J}$ so that $\tilde{\mu}$ is stable is a question of major interest, and will be addressed in Chapter 3.

ℓ-Step Lookahead

An important extension of one-step lookahead minimization is *ℓ-step lookahead*, whereby at a state x_k we minimize the cost of the first $\ell > 1$ stages with the future costs approximated by a function $\tilde{J}$ (see Fig. 2.2.1). This minimization yields a control $\tilde{u}_k$ and a sequence $\tilde{\mu}_{k+1}, \ldots, \tilde{\mu}_{k+\ell-1}$. The control $\tilde{u}_k$ is applied at x_k, and defines the ℓ-step lookahead policy $\tilde{\mu}$ via $\tilde{\mu}(x_k) = \tilde{u}_k$. The sequence $\tilde{\mu}_{k+1}, \ldots, \tilde{\mu}_{k+\ell-1}$ is discarded. Actually, we may view ℓ-step lookahead minimization as the special case of its one-step counterpart where the lookahead function is the optimal cost function of

---

† Note that the general theory of abstract DP is developed with the use of extended real-valued functions, and without the attainment of minimum assumption; see [Ber22a].

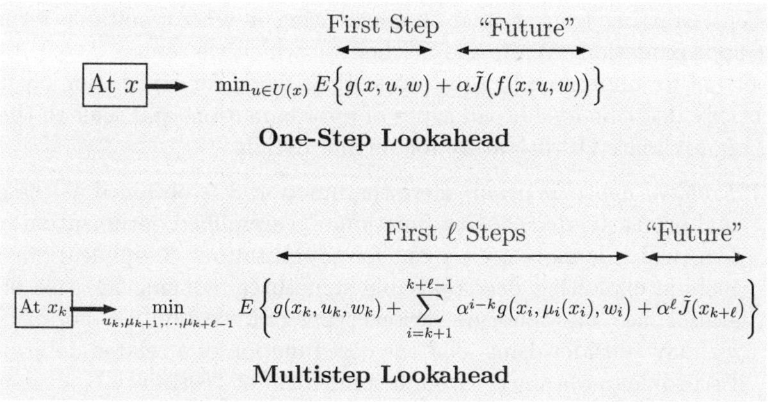

Sec. 2.2 Approximation in Value Space

$$\text{At } x \longrightarrow \min_{u \in U(x)} E\Big\{ g(x, u, w) + \alpha \tilde{J}\big(f(x, u, w)\big) \Big\}$$

One-Step Lookahead

$$\text{At } x_k \longrightarrow \min_{u_k, \mu_{k+1}, \ldots, \mu_{k+\ell-1}} E\Big\{ g(x_k, u_k, w_k) + \sum_{i=k+1}^{k+\ell-1} \alpha^{i-k} g\big(x_i, \mu_i(x_i), w_i\big) + \alpha^\ell \tilde{J}(x_{k+\ell}) \Big\}$$

Multistep Lookahead

Figure 2.2.1 Schematic illustration of approximation in value space with one-step and ℓ-step lookahead minimization. In the former case, the minimization yields at state x a control $\tilde{u}$, which defines the one-step lookahead policy $\tilde{\mu}$ via $\tilde{\mu}(x) = \tilde{u}$. In the latter case, the minimization yields a control $\tilde{u}_k$ and a sequence $\tilde{\mu}_{k+1}, \ldots, \tilde{\mu}_{k+\ell-1}$. The control $\tilde{u}_k$ is applied at x_k, and defines the ℓ-step lookahead policy $\tilde{\mu}$ via $\tilde{\mu}(x_k) = \tilde{u}_k$. The sequence $\tilde{\mu}_{k+1}, \ldots, \tilde{\mu}_{k+\ell-1}$ is discarded.

an $(\ell-1)$-stage DP problem with a terminal cost $\tilde{J}(x_{k+\ell})$ on the state $x_{k+\ell}$ obtained after $\ell - 1$ stages. In the next chapter, this will be interpreted as a step of Newton's method for solving Bellman's equation, starting from a function $\hat{J}$, which is an "improvement" over $\tilde{J}$. In particular, $\hat{J}$ is obtained from $\tilde{J}$ by applying $\ell - 1$ successive value iterations.

The motivation for ℓ-step lookahead minimization is that *by increasing the value of ℓ, we may require a less accurate approximation $\tilde{J}$ to obtain good performance*. Otherwise expressed, for the same quality of cost function approximation, better performance may be obtained as ℓ becomes larger. This will be explained visually in the next section, and is also supported by error bounds, given for example in the books [Ber19a], [Ber20a]. In particular, for AlphaZero chess, long multistep lookahead is critical for good on-line performance. *Another motivation for multistep lookahead is to enhance the stability properties of the generated on-line policy.* On the other hand, the multistep lookahead minimization problem is more time consuming that the one-step lookahead counterpart of Eq. (2.17).

Constructing Terminal Cost Approximations

A major issue in value space approximation is the construction of suitable approximate cost functions $\tilde{J}$. This can be done in many different ways, giving rise to some of the principal RL methods. For example, $\tilde{J}$ may be constructed with a sophisticated off-line training method, as discussed in Chapter 1, in connection with chess and backgammon. Alternatively, the approximate values $\tilde{J}(x)$ may be obtained on-line as needed with truncated rollout, by running an off-line obtained policy for a suitably large number of steps, starting from x, and supplementing it with a suitable terminal cost approximation. While the method by which we obtain $\tilde{J}$ will not be important for understanding the ideas of this work, for orientation purposes we briefly describe four broad types of approximation, and refer to the RL and approximate DP literature for further details:

(a) *Problem approximation*: Here the function $\tilde{J}$ is obtained as the optimal or nearly optimal cost function of a simplified optimization problem, which is more convenient for computation. Simplifications may include exploiting decomposable structure, reducing the size of the state space, and ignoring various types of uncertainties. For example we may consider using as $\tilde{J}$ the cost function of a related deterministic problem, obtained through some form of "certainty equivalence," thus allowing computation of $\tilde{J}$ by gradient-based optimal control methods or shortest path-type methods.

   A major type of problem approximation method is *aggregation*, which is described and analyzed in the books [Ber12], [Ber19a], and the papers [Ber18b], [Ber18c]. Aggregation provides a systematic procedure to simplify a given problem by grouping states together into a relatively small number of subsets, called aggregate states. The optimal cost function of the simpler aggregate problem is computed by exact DP methods, possibly involving the use of simulation. This cost function is then used to provide an approximation $\tilde{J}$ to the optimal cost function J^* of the original problem, using some form of interpolation.

(b) *On-line simulation*, as in rollout algorithms, where we use a suboptimal policy μ to compute on-line when needed the values $\tilde{J}(x)$ to be exactly or approximately equal to $J_\mu(x)$. The policy μ may be obtained by any method, e.g., one based on heuristic reasoning, or off-line training based on a more principled approach, such as approximate policy iteration or approximation in policy space. Note that while simulation is time-consuming, it is uniquely well-suited for the use of parallel computation. This may be an important consideration for the practical implementation of rollout algorithms, particularly for stochastic problems.

(c) *On-line approximate optimization*, such as model predictive control (MPC), which will be discussed in more detail later. This approach involves the solution of a suitably constructed ℓ-step version of the problem. It can be viewed as either approximation in value space with ℓ-step lookahead, or as a form of rollout algorithm.

(d) *Parametric cost approximation*, where $\tilde{J}$ is obtained from a given parametric class of functions $J(x,r)$, where r is a parameter vector, selected by a suitable algorithm. The parametric class typically involves prominent characteristics of x called *features*, which can be obtained either through insight into the problem at hand, or by using

Sec. 2.2   Approximation in Value Space   29

training data and some form of neural network.

We refer to the neurodynamic programming book by Bertsekas and Tsitsiklis [BeT96], and the RL book by Sutton and Barto [SuB18], as well as the large number of subsequent RL and approximate DP books, which provide specific examples of cost function approximation methods and associated training algorithms.

Let us also mention that for problems with special structure, the terminal cost approximation may be chosen so that the one-step lookahead minimization (2.17) is facilitated. In fact, in favorable circumstances, the lookahead minimization may be carried out in closed form. An example is when the control enters linearly the system equation and quadratically in the cost function, while the terminal cost approximation is quadratic.

From Off-Line Training to On-Line Play

Generally off-line training will produce either just a cost approximation (as in the case of TD-Gammon), or just a policy (as for example by some approximation in policy space/policy gradient approach), or both (as in the case of AlphaZero). We have already discussed in this section one-step lookahead and multistep lookahead schemes to implement on-line approximation in value space using $\tilde{J}$; cf. Fig. 2.2.1. Let us now consider some additional possibilities, some of which involve the use of a policy μ that has been obtained off-line (possibly in addition to a terminal cost approximation). Here are some of the main possibilities:

(a) *Given a policy μ that has been obtained off-line*, we may use as terminal cost approximation $\tilde{J}$ the cost function J_μ of the policy. For the case of one-step lookahead, this requires a policy evaluation operation, and can be done on-line, by computing (possibly by simulation) just the values of
$$E\big\{J_\mu\big(f(x_k, u_k, w_k)\big)\big\}$$
that are needed [cf. Eq. (2.17)]. For the case of ℓ-step lookahead, the values
$$E\big\{J_\mu(x_{k+\ell})\big\}$$
for all states $x_{k+\ell}$ that are reachable in ℓ steps starting from x_k are needed. This is the simplest form of rollout, and only requires the off-line construction of the policy μ.

(b) *Given a terminal cost approximation $\tilde{J}$ that has been obtained off-line*, we may use it on-line to compute a one-step or multistep lookahead policy $\tilde{\mu}$. In a more powerful version of this scheme, the policy $\tilde{\mu}$ can in turn be used for rollout as in (a) above. In a variation of this scheme, we may also use $\tilde{J}$ for truncated rollout, to approximate the tail end of the rollout process (an example of this is the rollout-based TD-Gammon algorithm discussed in Section 1.2).

(c) *Given a policy μ and a terminal cost approximation $\tilde{J}$*, we may use them together in a truncated rollout scheme, whereby the tail end of the rollout with μ is approximated using the cost approximation $\tilde{J}$. This is similar to the truncated rollout scheme noted in (b) above, except that the policy μ is computed off-line rather than on-line using $\tilde{J}$ and one-step or multistep lookahead as in (b).

The preceding three possibilities are the principal ones for using the results of off-line training within on-line play schemes. Naturally, there are variations where additional information is computed off-line to facilitate and/or expedite the on-line play algorithm. As an example, in MPC, in addition to a terminal cost approximation, a target tube may need to be computed off-line in order to guarantee that some state constraints can be satisfied on-line; see the discussion of MPC in Section 5.4. Other examples of this type will be noted in the context of specific applications.

Finally, let us note that while we have emphasized approximation in value space with cost function approximation, our discussion applies to Q-factor approximation, involving functions

$$\tilde{Q}(x, u) \approx E\Big\{g(x, u, w) + \alpha J^*\big(f(x, u, w)\big)\Big\}.$$

The corresponding one-step lookahead scheme has the form

$$\tilde{\mu}(x) \in \arg \min_{u \in U(x)} E\Big\{g(x,u,w) + \alpha \min_{u' \in U(f(x,u,w))} \tilde{Q}\big(f(x,u,w), u'\big)\Big\}; \quad (2.18)$$

cf. Eq. (2.17). The second term on the right in the above equation represents the cost function approximation

$$\tilde{J}\big(f(x,u,w)\big) = \min_{u' \in U(f(x,u,w))} \tilde{Q}\big(f(x,u,w), u'\big).$$

The use of Q-factors is common in the "model-free" case where a computer simulator is used to generate samples of w, and corresponding values of g and f. Then, having obtained $\tilde{Q}$ through off-line training, the one-step lookahead minimization in Eq. (2.18) must be performed on-line with the use of the simulator.

2.3 NOTES AND SOURCES

Our discussion of exact DP in this chapter has been brief since our focus in this book will be on approximate DP and RL. The author's DP textbooks [Ber12], [Ber17a] provide an extensive discussion of finite and infinite horizon exact DP, and its applications to discrete and continuous spaces problems, using a notation and style that is consistent with the present book. The author's paper [Ber17b] focuses on deterministic, nonnegative cost infinite horizon problems, and provides a convergence analysis of the value and policy iteration algorithms.

3

An Abstract View of Reinforcement Learning

Contents
3.1. Bellman Operators . . . . p. 32
3.2. Approximation in Value Space and Newton's Method . . p. 39
3.3. Region of Stability . . . . p. 46
3.4. Policy Iteration, Rollout, and Newton's Method . . . . . p. 50
3.5. How Sensitive is On-Line Play to the Off-Line Training Process? . . . . p. 58
3.6. Why Not Just Train a Policy Network and Use it Without On-Line Play? . . . . p. 60
3.7. Multiagent Problems and Multiagent Rollout . . . . . . p. 61
3.8. On-Line Simplified Policy Iteration . . . . p. 66
3.9. Exceptional Cases . . . . p. 72
3.10. Notes and Sources . . . . p. 79

In this chapter we will use geometric constructions to obtain insight into Bellman's equation, the value and policy iteration algorithms, approximation in value space, and some of the properties of the corresponding one-step or multistep lookahead policy $\tilde{\mu}$. To understand these constructions, we need an abstract notational framework that is based on the operators that are involved in the Bellman equations.

3.1 BELLMAN OPERATORS

We denote by TJ the function of x that appears in the right-hand side of Bellman's equation. Its value at state x is given by

$$(TJ)(x) = \min_{u \in U(x)} E\Big\{g(x,u,w) + \alpha J\big(f(x,u,w)\big)\Big\}, \qquad \text{for all } x. \quad (3.1)$$

Also for each policy μ, we introduce the corresponding function $T_\mu J$, which has value at x given by

$$(T_\mu J)(x) = E\Big\{g(x,\mu(x),w) + \alpha J\big(f(x,\mu(x),w)\big)\Big\}, \qquad \text{for all } x. \quad (3.2)$$

Thus T and T_μ can be viewed as operators (broadly referred to as the *Bellman operators*), which map functions J to other functions (TJ or $T_\mu J$, respectively).†

An important property of the operators T and T_μ is that they are *monotone*, in the sense that if J and J' are two functions of x such that

$$J(x) \geq J'(x), \qquad \text{for all } x,$$

then we have

$$(TJ)(x) \geq (TJ')(x), \qquad (T_\mu J)(x) \geq (T_\mu J')(x), \qquad \text{for all } x \text{ and } \mu. \quad (3.3)$$

This monotonicity property is evident from Eqs. (3.1) and (3.2), where the values of J are multiplied by nonnegative numbers.

---

† Within the context of this work, the functions J on which T and T_μ operate will be real-valued functions of x, which we denote by $J \in R(X)$. We will assume throughout that the expected values in Eqs. (3.1) and (3.2) are well-defined and finite when J is real-valued. This implies that $T_\mu J$ will also be real-valued functions of x. On the other hand $(TJ)(x)$ may take the value $-\infty$ because of the minimization in Eq. (3.1). We allow this possibility, although our illustrations will primarily depict the case where TJ is real-valued. Note that the general theory of abstract DP is developed with the use of extended real-valued functions; see [Ber22a].

Sec. 3.1 Bellman Operators

Another important property is that the Bellman operator T_μ is *linear*, in the sense that it has the form $T_\mu J = G + A_\mu J$, where $G \in R(X)$ is some function and $A_\mu : R(X) \mapsto R(X)$ is an operator such that for any functions J_1, J_2, and scalars γ_1, γ_2, we have†

$$A_\mu(\gamma_1 J_1 + \gamma_2 J_2) = \gamma_1 A_\mu J_1 + \gamma_2 A_\mu J_2.$$

Moreover, from the definitions (3.1) and (3.2), we have

$$(TJ)(x) = \min_{\mu \in \mathcal{M}} (T_\mu J)(x), \quad \text{for all } x,$$

where $\mathcal{M}$ is the set of stationary policies. This is true because for any policy μ, there is no coupling constraint between the controls $\mu(x)$ and $\mu(x')$ that correspond to two different states x and x'. It follows that $(TJ)(x)$ *is a concave function of J for every x* (the pointwise minimum of linear functions is a concave function). This will be important for our interpretation of one-step and multistep lookahead minimization as a Newton iteration for solving the Bellman equation $J = TJ$.

Example 3.1.1 (A Two-State and Two-Control Example)

Assume that there are two states 1 and 2, and two controls u and v. Consider the policy μ that applies control u at state 1 and control v at state 2. Then the operator T_μ takes the form

$$(T_\mu J)(1) = \sum_{y=1}^{2} p_{1y}(u)\big(g(1, u, y) + \alpha J(y)\big), \tag{3.4}$$

$$(T_\mu J)(2) = \sum_{y=1}^{2} p_{2y}(v)\big(g(2, v, y) + \alpha J(y)\big), \tag{3.5}$$

where $p_{xy}(u)$ and $p_{xy}(v)$ are the probabilities that the next state will be y, when the current state is x, and the control is u or v, respectively. Clearly, $(T_\mu J)(1)$ and $(T_\mu J)(2)$ are linear functions of J. Also the operator T of the Bellman equation $J = TJ$ takes the form

$$(TJ)(1) = \min\Bigg[\sum_{y=1}^{2} p_{1y}(u)\big(g(1, u, y) + \alpha J(y)\big), \\ \sum_{y=1}^{2} p_{1y}(v)\big(g(1, v, y) + \alpha J(y)\big)\Bigg], \tag{3.6}$$

---

† An operator T_μ with this property is often called "affine," but in this work we just call it "linear." Also we use abbreviated notation to express pointwise equalities and inequalities, so that we write $J = J'$ or $J \geq J'$ to express the fact that $J(x) = J'(x)$ or $J(x) \geq J'(x)$, for all x, respectively.

$$(TJ)(2) = \min\left[\sum_{y=1}^{2} p_{2y}(u)\big(g(2,u,y) + \alpha J(y)\big),\right.$$
$$\left.\sum_{y=1}^{2} p_{2y}(v)\big(g(2,v,y) + \alpha J(y)\big)\right]. \tag{3.7}$$

Thus, $(TJ)(1)$ and $(TJ)(2)$ are concave and piecewise linear as functions of the two-dimensional vector J (with two pieces; more generally, as many linear pieces as the number of controls). This concavity property holds in general since $(TJ)(x)$ is the minimum of a collection of linear functions of J, one for each $u \in U(x)$. Figure 3.1.1 illustrates $(T_\mu J)(1)$ for the cases where $\mu(1) = u$ and $\mu(1) = v$, $(T_\mu J)(2)$ for the cases where $\mu(2) = u$ and $\mu(2) = v$, $(TJ)(1)$, and $(TJ)(2)$, as functions of $J = \big(J(1), J(2)\big)$.

Critical properties from the DP point of view are whether T and T_μ have fixed points; equivalently, whether the Bellman equations $J = TJ$ and $J = T_\mu J$ have solutions within the class of real-valued functions, and whether the set of solutions includes J^* and J_μ, respectively. It may thus be important to verify that T or T_μ are contraction mappings. This is true for example in the benign case of discounted problems with bounded cost per stage. However, for undiscounted problems, asserting the contraction property of T or T_μ may be more complicated, and even impossible; the abstract DP book [Ber22a] deals extensively with such questions, and related issues regarding the solution sets of the Bellman equations.

Geometrical Interpretations

We will now interpret the Bellman operators geometrically, starting with T_μ. Figure 3.1.2 illustrates its form. Note here that the functions J and $T_\mu J$ are multidimensional. They have as many scalar components $J(x)$ and $(T_\mu J)(x)$, respectively, as there are states x, but they can only be shown projected onto one dimension. The function $T_\mu J$ for each policy μ is linear. The cost function J_μ satisfies $J_\mu = T_\mu J_\mu$, so it is obtained from the intersection of the graph of $T_\mu J$ and the 45 degree line, when J_μ is real-valued. Later we will interpret the case where J_μ is not real-valued as the system being unstable under μ [we have $J_\mu(x) = \infty$ for some initial states x].

The form of the Bellman operator T is illustrated in Fig. 3.1.3. Again the functions J, J^*, TJ, $T_\mu J$, etc, are multidimensional, but they are shown projected onto one dimension (alternatively they are illustrated for a system with a single state, plus possibly a termination state). The Bellman equation $J = TJ$ may have one or many real-valued solutions. It may also have no real-valued solution in exceptional situations, as we will discuss later (see Section 3.8). The figure assumes a unique real-valued solution of the Bellman equations $J = TJ$ and $J = T_\mu J$, which is true if T and T_μ are contraction mappings, as is the case for discounted problems with

Sec. 3.1   Bellman Operators   35

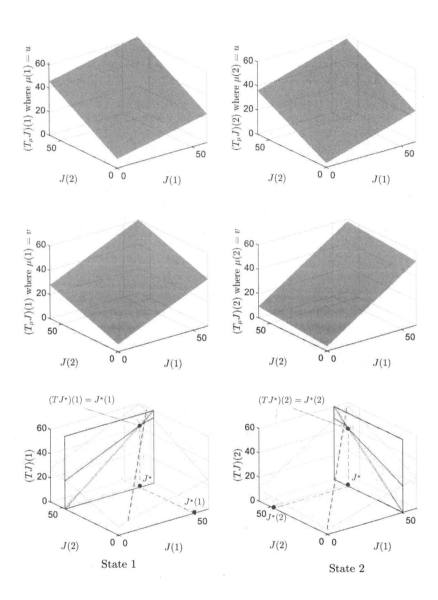

Figure 3.1.1 Geometric illustrations of the Bellman operators T_μ and T for states 1 and 2 in Example 3.1.1; cf. Eqs. (3.4)-(3.7). The problem's transition probabilities are: $p_{11}(u) = 0.3$, $p_{12}(u) = 0.7$, $p_{21}(u) = 0.4$, $p_{22}(u) = 0.6$, $p_{11}(v) = 0.6$, $p_{12}(v) = 0.4$, $p_{21}(v) = 0.9$, $p_{22}(v) = 0.1$. The stage costs are $g(1, u, 1) = 3$, $g(1, u, 2) = 10$, $g(2, u, 1) = 0$, $g(2, u, 2) = 6$, $g(1, v, 1) = 7$, $g(1, v, 2) = 5$, $g(2, v, 1) = 3$, $g(2, v, 2) = 12$. The discount factor is $\alpha = 0.9$, and the optimal costs are $J^*(1) = 50.59$ and $J^*(2) = 47.41$. The optimal policy is $\mu^*(1) = v$ and $\mu^*(2) = u$. The figure also shows two one-dimensional slices of T that are parallel to the $J(1)$ and $J(2)$ axes and pass through J^*.

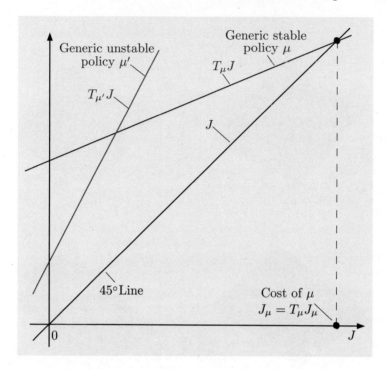

Figure 3.1.2 Geometric interpretation of the linear Bellman operator T_μ and the corresponding Bellman equation. The graph of T_μ is a plane in the space $R(X) \times R(X)$, and when projected on a one-dimensional plane that corresponds to a single state and passes through J_μ, it becomes a line. Then there are three cases:

(a) The line has slope less than 45 degrees, so it intersects the 45-degree line at a unique point, which is equal to J_μ, the solution of the Bellman equation $J = T_\mu J$. This is true if T_μ is a contraction mapping, as is the case for discounted problems with bounded cost per stage.

(b) The line has slope greater than 45 degrees. Then it intersects the 45-degree line at a unique point, which is a solution of the Bellman equation $J = T_\mu J$, but is not equal to J_μ. Then J_μ is not real-valued; we will call such μ *unstable* in Section 3.2.

(c) The line has slope exactly equal to 45 degrees. This is an exceptional case where the Bellman equation $J = T_\mu J$ has an infinite number of real-valued solutions or no real-valued solution at all; we will provide examples where this occurs in Section 3.8.

bounded cost per stage. Otherwise, these equations may have no solution or multiple solutions within the class of real-valued functions (see Section 3.8). The equation $J = TJ$ typically has J^* as a solution, but may have more than one solution in cases where either $\alpha = 1$, or $\alpha < 1$ and the cost per stage is unbounded.

Sec. 3.1 Bellman Operators

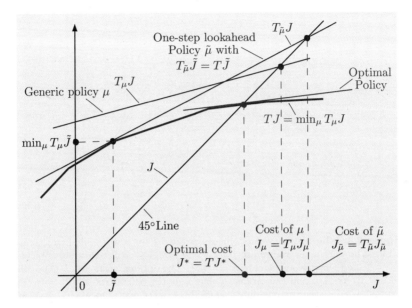

Figure 3.1.3 Geometric interpretation of the Bellman operator T, and the corresponding Bellman equation. For a fixed x, the function $(TJ)(x)$ can be written as $\min_\mu (T_\mu J)(x)$, so it is concave as a function of J. The optimal cost function J^* satisfies $J^* = TJ^*$, so it is obtained from the intersection of the graph of TJ and the 45 degree line shown, assuming J^* is real-valued.

Note that the graph of T lies below the graph of every operator T_μ, and is in fact obtained as the lower envelope of the graphs of T_μ as μ ranges over the set of policies $\mathcal{M}$. In particular, for any given function $\tilde{J}$, for every x, the value $(T\tilde{J})(x)$ is obtained by finding a support hyperplane/subgradient of the graph of the concave function $(TJ)(x)$ at $J = \tilde{J}$, as shown in the figure. This support hyperplane is defined by the control $\mu(x)$ of a policy $\tilde{\mu}$ that attains the minimum of $(T_\mu \tilde{J})(x)$ over μ:

$$\tilde{\mu}(x) \in \arg\min_{\mu \in \mathcal{M}} (T_\mu \tilde{J})(x)$$

(there may be multiple policies attaining this minimum, defining multiple support hyperplanes).

Example 3.1.2 (A Two-State and Infinite Controls Problem)

Let us consider the mapping T for a problem that involves two states, 1 and 2, but an infinite number of controls. In particular, the control space at both states is the unit interval, $U(1) = U(2) = [0, 1]$. Here $(TJ)(1)$ and $(TJ)(2)$ are given by

$$(TJ)(1) = \min_{u \in [0,1]} \left\{ g_1 + r_{11} u^2 + r_{12}(1-u)^2 + \alpha u J(1) + \alpha(1-u) J(2) \right\},$$

$$(TJ)(2) = \min_{u \in [0,1]} \left\{ g_2 + r_{21} u^2 + r_{22}(1-u)^2 + \alpha u J(1) + \alpha(1-u) J(2) \right\}.$$

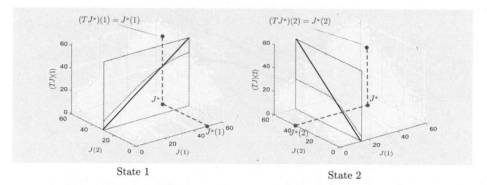

Figure 3.1.4 Illustration of the Bellman operator T for states 1 and 2 in Example 3.1.2. The parameter values are $g_1 = 5$, $g_2 = 3$, $r_{11} = 3$, $r_{12} = 15$, $r_{21} = 9$, $r_{22} = 1$, and the discount factor is $\alpha = 0.9$. The optimal costs are $J^*(1) = 49.7$ and $J^*(2) = 40.0$, and the optimal policy is $\mu^*(1) = 0.59$ and $\mu^*(2) = 0$. The figure also shows the two one-dimensional slices of the operators at $J(1) = 15$ and $J(2) = 30$ that are parallel to the $J(1)$ and $J(2)$ axes.

The control u at each state $x = 1, 2$ has the meaning of a probability that we must select at that state. In particular, we control the probabilities u and $(1-u)$ of moving to states $y = 1$ and $y = 2$, at a control cost that is quadratic in u and $(1-u)$, respectively. For this problem $(TJ)(1)$ and $(TJ)(2)$ can be calculated in closed form, so they are easy to plot and understand. They are piecewise quadratic, unlike the corresponding plots of Fig. 3.1.1, which are piecewise linear; see Fig. 3.1.4.

Visualization of Value Iteration

The operator notation simplifies algorithmic descriptions, derivations, and proofs related to DP. For example, we can write the VI algorithm in the compact form

$$J_{k+1} = TJ_k, \qquad k = 0, 1, \ldots,$$

as illustrated in Fig. 3.1.5. Moreover, the VI algorithm for a given policy μ can be written as

$$J_{k+1} = T_\mu J_k, \qquad k = 0, 1, \ldots,$$

and it can be similarly interpreted, except that the graph of the function $T_\mu J$ is linear. Also we will see shortly that there is a similarly compact description for the policy iteration algorithm.

To keep the presentation simple, we will focus our attention on the abstract DP framework as it applies to the optimal control problems of Section 2.1. In particular, we will assume without further mention that T and T_μ *have the monotonicity property (3.3), that $T_\mu J$ is linear for all μ, and*

Sec. 3.2   Approximation in Value Space and Newton's Method   39

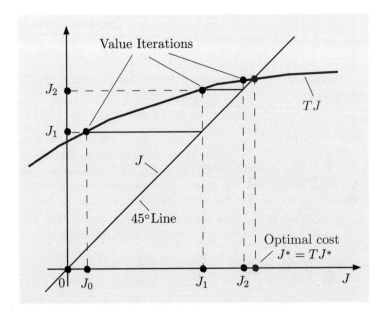

Figure 3.1.5 Geometric interpretation of the VI algorithm $J_{k+1} = TJ_k$, starting from some initial function J_0. Successive iterates are obtained through the staircase construction shown in the figure. The VI algorithm $J_{k+1} = T_\mu J_k$ for a given policy μ can be similarly interpreted, except that the graph of the function $T_\mu J$ is linear.

that (as a consequence) the component $(TJ)(x)$ is concave as a function of J for every state x. We note, however, that the abstract notation facilitates the extension of the infinite horizon DP theory to models beyond the ones that we discuss in this work. Such models include semi-Markov problems, minimax control problems, risk sensitive problems, Markov games, and others (see the DP textbook [Ber12], and the abstract DP monograph [Ber22a]).

3.2 APPROXIMATION IN VALUE SPACE AND NEWTON'S METHOD

Let us now consider approximation in value space and an abstract geometric interpretation, first provided in the author's book [Ber20a]. By using the operators T and T_μ, for a given $\tilde{J}$, a one-step lookahead policy $\tilde{\mu}$ is characterized by the equation $T_{\tilde{\mu}}\tilde{J} = T\tilde{J}$, or equivalently

$$\tilde{\mu}(x) \in \arg\min_{u \in U(x)} E\Big\{g(x, u, w) + \alpha \tilde{J}\big(f(x, u, w)\big)\Big\}, \qquad (3.8)$$

as in Fig. 3.2.1. Furthermore, this equation implies that the graph of $T_{\tilde{\mu}}J$ just touches the graph of TJ at $\tilde{J}$, as shown in the figure.

In mathematical terms, for each state $x \in X$, the hyperplane $H_{\tilde{\mu}}(x) \in R(X) \times \Re$

$$H_{\tilde{\mu}}(x) = \{(J, \xi) \mid (T_{\tilde{\mu}} J)(x) = \xi\}, \qquad (3.9)$$

supports from above the hypograph of the concave function $(TJ)(x)$, i.e., the convex set

$$\{(J, \xi) \mid (TJ)(x) \geq \xi\}.$$

The point of support is $\big(\tilde{J}, (T_{\tilde{\mu}} \tilde{J})(x)\big)$, and relates the function $\tilde{J}$ with the corresponding one-step lookahead minimization policy $\tilde{\mu}$, the one that satisfies $T_{\tilde{\mu}} \tilde{J} = T \tilde{J}$. The hyperplane $H_{\tilde{\mu}}(x)$ of Eq. (3.9) defines a subgradient of $(TJ)(x)$ at $\tilde{J}$. Note that the one-step lookahead policy $\tilde{\mu}$ need not be unique, since T need not be differentiable, so there may be multiple hyperplanes of support at $\tilde{J}$. Still this construction shows that the linear operator $T_{\tilde{\mu}}$ is a linearization of the operator T at the point $\tilde{J}$ (pointwise for each x).

Equivalently, for every $x \in X$, the linear scalar equation $J(x) = (T_{\tilde{\mu}} J)(x)$ is a linearization of the nonlinear equation $J(x) = (TJ)(x)$ at the point $\tilde{J}$. Consequently, the linear operator equation $J = T_{\tilde{\mu}} J$ is a linearization of the equation $J = TJ$ at $\tilde{J}$, and its solution, $J_{\tilde{\mu}}$, can be viewed as the result of a Newton iteration at the point $\tilde{J}$ (here we adopt an expanded view of the Newton iteration that applies to possibly nondifferentiable fixed point equations; see the Appendix). In summary, *the Newton iterate at $\tilde{J}$ is $J_{\tilde{\mu}}$, the solution of the linearized equation $J = T_{\tilde{\mu}} J$*.†

---

† The classical Newton's method for solving a fixed point problem of the form $y = G(y)$, where y is an n-dimensional vector, operates as follows: At the current iterate y_k, we linearize G and find the solution y_{k+1} of the corresponding linear fixed point problem. Assuming G is differentiable, the linearization is obtained by using a first order Taylor expansion:

$$y_{k+1} = G(y_k) + \frac{\partial G(y_k)}{\partial y}(y_{k+1} - y_k),$$

where $\partial G(y_k)/\partial y$ is the $n \times n$ Jacobian matrix of G evaluated at the vector y_k. The most commonly given convergence rate property of Newton's method is *quadratic convergence*. It states that near the solution y^*, we have

$$\|y_{k+1} - y^*\| = O\big(\|y_k - y^*\|^2\big),$$

where $\|\cdot\|$ is the Euclidean norm, and holds assuming the Jacobian matrix exists, is invertible, and is Lipschitz continuous (see the books by Ortega and Rheinboldt [OrR70], and by the author [Ber16], Section 1.4).

There are well-studied extensions of Newton's method that are based on solving a linearized system at the current iterate, but relax the differentiability requirement through alternative requirements of piecewise differentiability, B-differentiability, and semi-smoothness, while maintaining the superlinear convergence property of the method. In particular, the quadratic rate of convergence

Sec. 3.2   Approximation in Value Space and Newton's Method   41

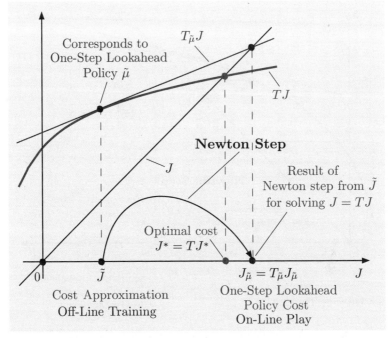

Figure 3.2.1 Geometric interpretation of approximation in value space and the one-step lookahead policy $\tilde{\mu}$ as a step of Newton's method [cf. Eq. (3.8)]. Given $\tilde{J}$, we find a policy $\tilde{\mu}$ that attains the minimum in the relation

$$T\tilde{J} = \min_{\mu} T_{\mu}\tilde{J}.$$

This policy satisfies $T\tilde{J} = T_{\tilde{\mu}}\tilde{J}$, so the graph of TJ and $T_{\tilde{\mu}}J$ touch at $\tilde{J}$, as shown. It may not be unique. Because TJ has concave components, the equation $J = T_{\tilde{\mu}}J$ is the linearization of the equation $J = TJ$ at $\tilde{J}$ [for each x, the hyperplane $H_{\tilde{\mu}}(x)$ of Eq. (3.9) defines a subgradient of $(TJ)(x)$ at $\tilde{J}$]. The linearized equation is solved at the typical step of Newton's method to provide the next iterate, which is just $J_{\tilde{\mu}}$.

The structure of the Bellman operators (3.1) and (3.2), with their monotonicity and concavity properties, tends to enhance the convergence and the rate of convergence properties of Newton's method, even in the absence of differentiability, as evidenced by the favorable Newton-related convergence analysis of PI, and the extensive favorable experience with rollout, PI, and MPC. In fact, the role of monotonicity and concavity in affecting the convergence properties of Newton's method has been addressed

---

result for differentiable G of Prop. 1.4.1 of the book [Ber16] admits a straightforward and intuitive extension to piecewise differentiable G, given in the paper [KoS86]; see the Appendix, which contains references to the literature.

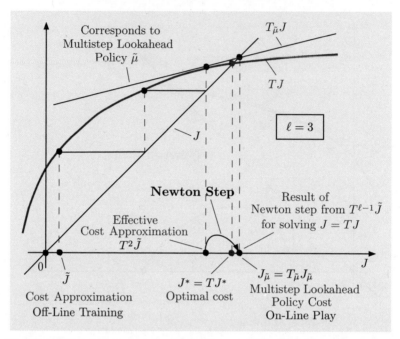

Figure 3.2.2 Geometric interpretation of approximation in value space with ℓ-step lookahead (in this figure $\ell = 3$). It is the same as approximation in value space with one-step lookahead using $T^{\ell-1}\tilde{J}$ as cost approximation. It can be viewed as a Newton step at the point $T^{\ell-1}\tilde{J}$, the result of $\ell - 1$ value iterations applied to $\tilde{J}$. Note that as ℓ increases the cost function $J_{\tilde{\mu}}$ of the ℓ-step lookahead policy $\tilde{\mu}$ approaches more closely the optimal J^*, and that $\lim_{\ell \to \infty} J_{\tilde{\mu}} = J^*$.

in the mathematical literature.†

As noted earlier, approximation in value space with ℓ-step lookahead using $\tilde{J}$ is the same as approximation in value space with one-step lookahead using the $(\ell-1)$-fold operation of T on $\tilde{J}$, $T^{\ell-1}\tilde{J}$. Thus it can be interpreted as a Newton step starting from $T^{\ell-1}\tilde{J}$, the result of $\ell - 1$ value iterations applied to $\tilde{J}$. This is illustrated in Fig. 3.2.2.‡

---

† See the papers by Ortega and Rheinboldt [OrR67], and Vandergraft [Van67], the books by Ortega and Rheinboldt [OrR70], and Argyros [Arg08], and the references cited there. In this connection, it is worth noting that in the case of Markov games, where the concavity property does not hold, the PI method may oscillate, as shown by Pollatschek and Avi-Itzhak [PoA69], and needs to be modified to restore its global convergence; see the author's paper [Ber21c], and the references cited there.

‡ We note that several variants of Newton's method that involve combinations of first-order iterative methods, such as the Gauss-Seidel and Jacobi algorithms, and Newton's method, are well-known in numerical analysis. They belong to the general family of *Newton-SOR methods* (SOR stands for "succes-

Sec. 3.2   Approximation in Value Space and Newton's Method   43

Let us also note that ℓ-step lookahead minimization involves ℓ successive VI iterations, but *only the first of these iterations has a Newton step interpretation*. As an example, consider two-step lookahead minimization with a terminal cost approximation $\tilde{J}$. The second step minimization is a VI that starts from $\tilde{J}$ to produce $T\tilde{J}$. The first step minimization is a VI that starts from $T\tilde{J}$ to produce $T^2\tilde{J}$, but it also does something else that is more significant: *It produces a two-step lookahead minimization policy $\tilde{\mu}$ through $T_{\tilde{\mu}}(T\tilde{J}) = T(T\tilde{J})$, and the step from $T\tilde{J}$ to $J_{\tilde{\mu}}$ (the cost function of $\tilde{\mu}$) is the Newton step*. Thus, there is only one policy produced (i.e., $\tilde{\mu}$) and only one Newton step (from $T\tilde{J}$ to $J_{\tilde{\mu}}$). In the case of one-step lookahead minimization, the Newton step starts from $\tilde{J}$ and ends at $J_{\tilde{\mu}}$. Similarly, in the case of ℓ-step lookahead minimization, the first step of the lookahead is the Newton step (from $T^{\ell-1}\tilde{J}$ to $J_{\tilde{\mu}}$), and whatever follows the first step of the lookahead is preparation for the Newton step.

Finally, it is worth mentioning that the approximation in value space algorithm computes $J_{\tilde{\mu}}$ differently than both the PI method and the classical form of Newton's method. It does not explicitly compute any values of $J_{\tilde{\mu}}$; instead, the control is applied to the system and the cost is accumulated accordingly. Thus the values of $J_{\tilde{\mu}}$ are implicitly computed only for those x that are encountered in the system trajectory that is generated on-line.

Certainty Equivalent Approximations and the Newton Step

We noted earlier that for stochastic DP problems, ℓ-step lookahead can be computationally expensive, because the lookahead graph expands fast as ℓ increases, due to the stochastic character of the problem. The *certainty equivalence approach* is an important approximation idea for dealing with this difficulty. In the classical form of this approach, some or all of the stochastic disturbances w_k are replaced by some deterministic quantities, such as their expected values. Then a policy is computed off-line for the resulting deterministic problem, and it is used on-line for the actual stochastic problem.

The certainty equivalence approach can also be used to expedite the computations of the ℓ-step lookahead minimization. One way to do this is to simply replace each of the uncertain ℓ quantities $w_k, w_{k+1}, \ldots, w_{k+\ell-1}$ by a deterministic value $\overline{w}$. Conceptually, this replaces the Bellman operators T and T_μ,

$$(TJ)(x) = \min_{u \in U(x)} E\Big\{g(x,u,w) + \alpha J\big(f(x,u,w)\big)\Big\},$$

---

sive over-relaxation"); see the book by Ortega and Rheinboldt [OrR70] (Section 13.4). Their convergence rate is superlinear, similar to Newton's method, as long as they involve a pure Newton step, along with the first-order steps.

$$(T_\mu J)(x) = E\big\{g(x,\mu(x),w) + \alpha J(f(x,\mu(x),w))\big\},$$

[cf. Eqs. (3.1) and (3.2)] with deterministic operators $\overline{T}$ and $\overline{T}_\mu$, given by

$$(\overline{T}J)(x) = \min_{u\in U(x)}\big[g(x,u,\overline{w}) + \alpha J(f(x,u,\overline{w}))\big],$$

$$(\overline{T}_\mu J)(x) = g(x,\mu(x),\overline{w}) + \alpha J(f(x,\mu(x),\overline{w})).$$

The resulting ℓ-step lookahead minimization then becomes simpler; for example, in the case of a finite control space problem, it is a deterministic shortest path computation, involving an acyclic ℓ-stage graph that expands at each stage by a factor n, where n is the size of the control space. However, this approach yields a policy $\overline{\mu}$ such that

$$\overline{T}_{\overline{\mu}}(\overline{T}^{\ell-1}\tilde{J}) = \overline{T}(\overline{T}^{\ell-1}\tilde{J}),$$

and the cost function $J_{\overline{\mu}}$ of this policy is generated by a Newton step, which aims to find a fixed point of $\overline{T}$ (not T), starting from $\overline{T}^{\ell-1}\tilde{J}$. Thus the Newton step now aims at a fixed point of $\overline{T}$, which is not equal to J^*. As a result the benefit of the Newton step is lost to a great extent.

Still, we may largely correct this difficulty, while retaining substantial simplification, by using certainty equivalence for *only the last $\ell-1$ stages* of the ℓ-step lookahead. This can be done with an ℓ-step lookahead scheme whereby only the uncertain quantities $w_{k+1},\ldots,w_{k+\ell-1}$ are replaced by a deterministic value $\overline{w}$, while w_k is treated as a stochastic quantity, as first proposed in the paper by Bertsekas and Castañon [BeC99]. In this way we obtain a policy $\overline{\mu}$ such that

$$T_{\overline{\mu}}(\overline{T}^{\ell-1}\tilde{J}) = T(\overline{T}^{\ell-1}\tilde{J}).$$

The cost function $J_{\overline{\mu}}$ of this policy is then generated by a Newton step, which aims to find a fixed point of T (not $\overline{T}$), starting again from $\overline{T}^{\ell-1}\tilde{J}$. Thus the benefit of the fast convergence of Newton's method is restored. In fact based on insights derived from this Newton step interpretation, it appears that the performance penalty for making the last $\ell-1$ stages of the ℓ-step lookahead deterministic is minimal when $\overline{T}^{\ell-1}\tilde{J}$ is "near" J^*. At the same time the ℓ-step minimization $T(\overline{T}^{\ell-1}\tilde{J})$ involves only one stochastic step, the first one, and hence potentially a much "thinner" lookahead graph, than the one corresponding to the ℓ-step minimization $T^\ell \tilde{J}$, which does not involve any certainty equivalence-type approximations.

The preceding discussion also points to a more general approximation idea for dealing with the onerous computational requirements of long multistep lookahead minimization. We may approximate the tail $(\ell-1)$-step portion $T^{\ell-1}\tilde{J}$ of the ℓ-step lookahead minimization with any simplified

Sec. 3.2   Approximation in Value Space and Newton's Method

calculation that produces an approximation $\hat{J} \approx T^{\ell-1}\tilde{J}$, and then obtain the lookahead policy $\tilde{\mu}$ using the minimization

$$T_{\tilde{\mu}}\hat{J} = T\hat{J}.$$

This type of simplification will still involve a Newton step (from $\hat{J}$ to $J_{\tilde{\mu}}$), and benefit from the corresponding fast convergence property.

Local and Global Performance Estimates Compared

The preceding Newton step interpretation of the move from $\tilde{J}$ (the terminal cost function approximation) to $J_{\tilde{\mu}}$ (the cost function of the lookahead policy $\tilde{\mu}$) suggests a superlinear performance estimate

$$\max_{x}\big|J_{\tilde{\mu}}(x) - J^{*}(x)\big| = o\Big(\max_{x}\big|\tilde{J}(x) - J^{*}(x)\big|\Big).$$

However, this estimate is *local* in character. It is meaningful only when $\tilde{J}$ is "close" to J^*. When $\tilde{J}$ is far from J^*, the difference $\max_x\big|J_{\tilde{\mu}}(x) - J^*(x)\big|$ may be large and even infinite when $\tilde{\mu}$ is unstable (see the discussion in the next section).

There are global estimates for the difference

$$\max_{x}\big|J_{\tilde{\mu}}(x) - J^{*}(x)\big|$$

for several types of problems, including the upper bound

$$\max_{x}\big|J_{\tilde{\mu}}(x) - J^{*}(x)\big| \le \frac{2\alpha^{\ell}}{1-\alpha}\max_{x}\big|\tilde{J}(x) - J^{*}(x)\big|$$

for ℓ-step lookahead, and α-discounted problems where all the Bellman operators T_μ are contraction mappings; see the neurodynamic programming book [BeT96] (Section 6.1, Prop. 6.1), or the RL book [Ber20a] (Section 5.4, Prop. 5.4.1). These books also contain other related global estimates, which hold for all $\tilde{J}$, both close and far from J^*. However, these global estimates tend to be overly conservative and not representative of the performance of approximation in value space schemes when $\tilde{J}$ is near J^*. For example, for finite spaces α-discounted MDP, $\tilde{\mu}$ can be shown to be optimal when $\max_x\big|\tilde{J}(x) - J^*(x)\big|$ is sufficiently small; this can also be seen from the fact that the components $(TJ)(x)$ of the Bellman operator are not only concave but also piecewise linear, so Newton's method converges finitely. For a further comparative discussion of local and global error bounds, we refer to Appendix A.

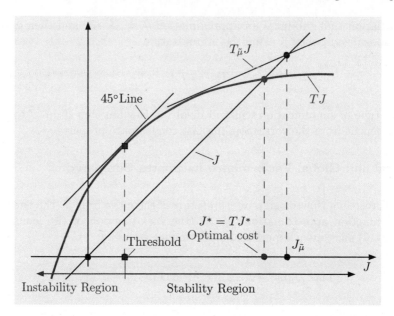

Figure 3.3.1 Illustration of the regions of stability and instability for approximation in value space with one-step lookahead. The stability region is the set of all $\tilde{J}$ such that the policy $\tilde{\mu}$ obtained from the one-step lookahead minimization $T_{\tilde{\mu}}\tilde{J} = T\tilde{J}$ satisfies $J_{\tilde{\mu}}(x) < \infty$ for all x.

3.3 REGION OF STABILITY

For any control system design method, the stability of the policy obtained is of paramount importance. It is thus essential to investigate and verify the stability of controllers obtained through approximation in value space schemes. Historically, there have been several proposed definitions of stability in control theory. Within the context of this work, our focus on stability issues will be for problems with a termination state t, which is cost-free, and with a cost per stage that is positive outside the termination state, such as the undiscounted positive cost deterministic problem introduced earlier (cf. Section 2.1). Moreover, it is best for our purposes to adopt an optimization-based definition. In particular, we say that a policy μ is *unstable* if $J_\mu(x) = \infty$ for some states x. Equivalently, we say that the policy μ *stable* if $J_\mu(x) < \infty$ for all states x. This definition has the advantage that it applies to general state and control spaces. Naturally, it can be made more specific in particular problem instances.†

---

† For the undiscounted positive cost deterministic problem introduced earlier (cf. Section 2.1), it can be shown that if a policy μ is stable, then J_μ is the "smallest" solution of the Bellman equation $J = T_\mu J$ within the class of nonnegative real-valued functions, and under mild assumptions it is the unique solution of

Sec. 3.3   Region of Stability   47

In the context of approximation in value space we are interested in the *region of stability*, which is the set of cost function approximations $\tilde{J} \in R(X)$ for which the corresponding one-step or multistep lookahead policies $\tilde{\mu}$ are stable. For discounted problems with bounded cost per stage, all policies have real-valued cost functions, so questions of stability do not arise. In general, however, the region of stability may be a strict subset of the set of real-valued functions; this will be illustrated later for the undiscounted deterministic case of the linear quadratic problem of Section 2.1 (cf. Example 2.1.1). Figure 3.3.1 illustrates the region of stability for approximation in value space with one-step lookahead.

An interesting observation from Fig. 3.3.1 is that if $\tilde{J}$ does not belong to the region of stability and $\tilde{\mu}$ is a corresponding one-step lookahead unstable policy, the Bellman equation $J = T_{\tilde{\mu}} J$ may have real-valued solutions. However, these solutions will not be equal to $J_{\tilde{\mu}}$, as this would violate the definition of region of stability. Generally, if T_μ is not a contraction mapping, T_μ may have real-valued fixed points, none of which is equal to J_μ.

Figure 3.3.2 illustrates the region of stability for the case of multistep lookahead minimization. The insights from this figure are similar to the one-step lookahead case of Fig. 3.3.1. However, the figure indicates that *the region of stability of the ℓ-step lookahead controller $\tilde{\mu}$ depends on ℓ, and tends to become larger as ℓ increases*. The reason is that ℓ-step lookahead with terminal cost $\tilde{J}$ is equivalent to one-step lookahead with terminal cost $T^{\ell-1} \tilde{J}$, which tends to be closer to the optimal cost function J^* than $\tilde{J}$ (assuming convergence of the VI method).

How Can We Obtain Function Approximations $\tilde{J}$ Within the Region of Stability?

Naturally, identifying and obtaining cost function approximations $\tilde{J}$ that lie within the region of stability with either one-step or multistep lookahead is very important within our context. We will focus on this question for the special case where the expected cost per stage is nonnegative

$$E\{g(x, u, w)\} \geq 0, \qquad \text{for all } x, \, u \in U(x),$$

and assume that J^* is real-valued. This is the case of most interest in model predictive control, but also arises in other problems of interest, including stochastic shortest path problems that involve a termination state.

From Fig. 3.3.2 it can be conjectured that if the sequence $\{T^k \tilde{J}\}$ generated by the VI algorithm converges to J^* for all $\tilde{J}$ such that $0 \leq \tilde{J} \leq J^*$

$J = T_\mu J$ within the class of nonnegative real-valued functions J with $J(t) = 0$; see the author's paper [Ber17b]. Moreover, if μ is unstable, then the Bellman equation $J = T_\mu J$ has no solution within the class of nonnegative real-valued functions.

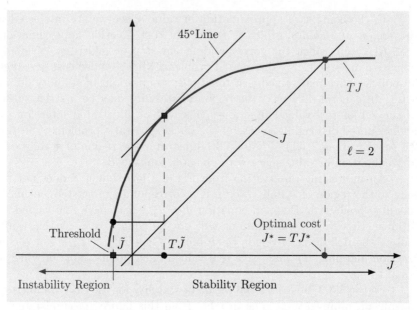

Figure 3.3.2 Illustration of the regions of stability and instability for approximation in value space with ℓ-step lookahead minimization. The stability region is the set of all $\tilde{J}$ for which the policy $\tilde{\mu}$ such that $T^\ell \tilde{J} = T_{\tilde{\mu}} T^{\ell-1} \tilde{J}$ satisfies $J_{\tilde{\mu}}(x) < \infty$ for all x (the figure shows the case $\ell = 2$). The region of instability tends to be reduced as ℓ increases.

(which is true under very general conditions; see [Ber12], [Ber22a]), then $T^{\ell-1}\tilde{J}$ belongs to the region of stability for sufficiently large ℓ. Related ideas have been discussed in the adaptive DP literature by Liu and his collaborators [HWL21], [LXZ21], [WLL16], and by Heydari [Hey17], [Hey18], who provide extensive references; see also Winnicki et al. [WLL21]. We will revisit this issue in the context of linear quadratic problems. This conjecture is generally true, but requires that, in addition to J^*, all functions $\tilde{J}$ within a neighborhood of J^* belong to the region of stability. Our subsequent discussion will aim to address this difficulty.

An important fact for our nonnegative cost problem context is that *the region of stability includes all real-valued nonnegative functions $\tilde{J}$ such that*

$$T\tilde{J} \leq \tilde{J}. \tag{3.10}$$

Indeed if $\tilde{\mu}$ is the corresponding one-step lookahead policy, we have

$$T_{\tilde{\mu}}\tilde{J} = T\tilde{J} \leq \tilde{J},$$

and from a well-known result on nonnegative cost infinite horizon problems [see [Ber12], Prop. 4.1.4(a)], it follows that

$$J_{\tilde{\mu}} \leq \tilde{J};$$

Sec. 3.3 Region of Stability

(the proof argument is that if $T_{\tilde{\mu}}\tilde{J} \leq \tilde{J}$ then $T_{\tilde{\mu}}^{k+1}\tilde{J} \leq T_{\tilde{\mu}}^k \tilde{J}$ for all k, so, using also the fact $0 \leq \tilde{J}$, the limit of $T_{\tilde{\mu}}^k \tilde{J}$, call it J_∞, satisfies $J_{\tilde{\mu}} \leq J_\infty \leq \tilde{J}$). Thus if $\tilde{J}$ is nonnegative and real-valued, $J_{\tilde{\mu}}$ is also real-valued, so $\tilde{\mu}$ is stable. It follows that $\tilde{J}$ belongs to the region of stability. This is a known result in specific contexts, such as MPC (see the book by Rawlings, Mayne, and Diehl [RMD17], Section 2.4, which contains extensive references to prior work on stability issues).

An important special case where the condition $T\tilde{J} \leq \tilde{J}$ is satisfied is when $\tilde{J}$ is the cost function of a stable policy, i.e., $\tilde{J} = J_\mu$. Then we have that J_μ is real-valued and satisfies $T_\mu J_\mu = J_\mu$, so it follows that $TJ_\mu \leq J_\mu$. This case relates to the rollout algorithm and shows that *rollout with a stable policy yields a stable lookahead policy*. It also suggests that if μ is stable, then $T_\mu^m \tilde{J}$ belongs to the region of stability for sufficiently large m.

Besides J_μ, with stable μ, and J^*, there are other interesting functions $\tilde{J}$ satisfying the stability condition $T\tilde{J} \leq \tilde{J}$. In particular, let β be a scalar with $\beta > 1$, and for a stable policy μ, consider the *β-amplified operator* $T_{\mu,\beta}$ defined by

$$(T_{\mu,\beta}J)(x) = E\Big\{\beta g\big(x,\mu(x),w\big) + \alpha J\big(f(x,\mu(x),w)\big)\Big\}, \qquad \text{for all } x.$$

Then it can be seen that the function

$$J_{\mu,\beta} = \beta J_\mu$$

is a fixed point of $T_{\mu,\beta}$ and satisfies $TJ_{\mu,\beta} \leq J_{\mu,\beta}$. This follows by writing

$$J_{\mu,\beta} = T_{\mu,\beta} J_{\mu,\beta} \geq T_\mu J_{\mu,\beta} \geq T J_{\mu,\beta}. \tag{3.11}$$

Thus $J_{\mu,\beta}$ lies within the region of stability, and lies "further to the right" of J_μ. Thus we may conjecture that it can be more reliably approximated by $T_{\mu,\beta}^m \tilde{J}$ than J_μ is approximated by $T_\mu^m \tilde{J}$ in the context of m-step truncated rollout.

To illustrate this fact, consider a stable policy μ, and assume that the expected cost per stage at states other than a termination state t (if one exists) is bounded away from 0, i.e.,

$$C = \min_{x \neq t} E\Big\{g\big(x,\mu(x),w\big)\Big\} > 0.$$

Then we claim that given a scalar $\beta > 1$, any function $\hat{J} \in R(X)$ with $\hat{J}(t) = 0$, that satisfies

$$\max_x \big|\hat{J}(x) - J_{\mu,\beta}(x)\big| \leq \delta, \qquad \text{for all } x, \tag{3.12}$$

where

$$\delta = \frac{(\beta-1)C}{1+\alpha},$$

also satisfies the stability condition $T\hat{J} \leq \hat{J}$. From this it follows that *for a given nonnegative and real-valued $\tilde{J}$, and for sufficiently large m, so that the function $\hat{J} = T_{\mu,\beta}^m \tilde{J}$ satisfies Eq. (3.12), we have that $\hat{J}$ lies within the region of stability.*

To see this, note that for all $x \neq t$, we have

$$J_{\mu,\beta}(x) = \beta E\big\{g(x,\mu(x),w)\big\} + \alpha E\big\{J_{\mu,\beta}(f(x,\mu(x),w))\big\},$$

so that by using Eq. (3.12), we have

$$\hat{J}(x) + \delta \geq \beta E\big\{g(x,\mu(x),w)\big\} + \alpha E\big\{\hat{J}(f(x,\mu(x),w))\big\} - \alpha\delta.$$

It follows that

$$\begin{aligned}
\hat{J}(x) &\geq E\big\{g(x,\mu(x),w)\big\} + \alpha E\big\{\hat{J}(f(x,\mu(x),w))\big\} \\
&\quad + (\beta-1)E\big\{g(x,\mu(x),w)\big\} - (1+\alpha)\delta \\
&\geq E\big\{g(x,\mu(x),w)\big\} + \alpha E\big\{\hat{J}(f(x,\mu(x),w))\big\} + (\beta-1)C - (1+\alpha)\delta \\
&= (T_\mu \hat{J})(x) \\
&\geq (T\hat{J})(x),
\end{aligned}$$

so the stability condition $T\hat{J} \leq \hat{J}$ is satisfied.

Similarly the function

$$J_\beta^* = \beta J^*$$

is a fixed point of the operator T_β defined by

$$(T_\beta J)(x) = \min_{u \in U(x)} E\big\{\beta g(x,u,w) + \alpha J(f(x,u,w))\big\}, \qquad \text{for all } x.$$

It can be seen, using an argument similar to Eq. (3.11), that J_β^* satisfies $TJ_\beta^* \leq J_\beta^*$, so it lies within the region of stability. Furthermore, similar to the case of truncated rollout discussed earlier, we may conjecture that J_β^* can be more reliably approximated by $T_\beta^{\ell-1}\tilde{J}$ than J^* is approximated by $T^{\ell-1}\tilde{J}$ in the context of ℓ-step lookahead.

3.4 POLICY ITERATION, ROLLOUT, AND NEWTON'S METHOD

Another major class of infinite horizon algorithms is based on *policy iteration* (PI for short), which involves the repeated use of policy improvement, in analogy with the AlphaZero/TD-Gammon off-line training algorithms,

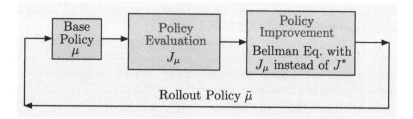

Figure 3.4.1 Schematic illustration of PI as repeated rollout. It generates a sequence of policies, with each policy μ in the sequence being the base policy that generates the next policy $\tilde{\mu}$ in the sequence as the corresponding rollout policy.

described in Chapter 1. Each iteration of the PI algorithm starts with a stable policy (which we call *current* or *base* policy), and generates another stable policy (which we call *new* or *rollout* policy, respectively). For the infinite horizon problem of Section 2.1, given the base policy μ, the iteration consists of two phases (see Fig. 3.4.1):

(a) *Policy evaluation*, which computes the cost function J_μ. One possibility is to solve the corresponding Bellman equation

$$J_\mu(x) = E\big\{g(x,\mu(x),w) + \alpha J_\mu\big(f(x,\mu(x),w)\big)\big\}, \qquad \text{for all } x. \tag{3.13}$$

However, the value $J_\mu(x)$ for any x can also be computed by Monte Carlo simulation, by averaging over many randomly generated trajectories the cost of the policy starting from x. Other, more sophisticated possibilities include the use of specialized simulation-based methods, such as *temporal difference methods*, for which there is extensive literature (see e.g., the books [BeT96], [SuB98], [Ber12]).

(b) *Policy improvement*, which computes the rollout policy $\tilde{\mu}$ using the one-step lookahead minimization

$$\tilde{\mu}(x) \in \arg\min_{u \in U(x)} E\big\{g(x,u,w) + \alpha J_\mu\big(f(x,u,w)\big)\big\}, \qquad \text{for all } x. \tag{3.14}$$

It is generally expected (and can be proved under mild conditions) that the rollout policy is improved in the sense that

$$J_{\tilde{\mu}}(x) \leq J_\mu(x), \qquad \text{for all } x.$$

Proofs of this fact in a variety of contexts can be found in most DP books, including the author's [Ber12], [Ber18a], [Ber19a], [Ber20a], [Ber22a].

Thus PI generates a sequence of stable policies $\{\mu^k\}$, by obtaining μ^{k+1} through a policy improvement operation using J_{μ^k} in place of J_μ

in Eq. (3.14), which is obtained through policy evaluation of the preceding policy μ^k using Eq. (3.13). It is well known that (exact) PI has solid convergence properties; see the DP textbooks cited earlier, as well as the author's RL book [Ber19a]. These properties hold even when the method is implemented (with appropriate modifications) in unconventional computing environments, involving asynchronous distributed computation, as shown in a series of papers by Bertsekas and Yu [BeY10], [BeY12], [YuB13].

In terms of our abstract notation, the PI algorithm can be written in a compact form. For the generated policy sequence $\{\mu^k\}$, the policy evaluation phase obtains J_{μ^k} from the equation

$$J_{\mu^k} = T_{\mu^k} J_{\mu^k}, \tag{3.15}$$

while the policy improvement phase obtains μ^{k+1} through the equation

$$T_{\mu^{k+1}} J_{\mu^k} = T J_{\mu^k}. \tag{3.16}$$

As Fig. 3.4.2 illustrates, PI can be viewed as Newton's method for solving the Bellman equation in the function space of cost functions J. In particular, *the policy improvement Eq. (3.16) is the Newton step starting from J_{μ^k}, and yields μ^{k+1} as the corresponding one-step lookahead/rollout policy.* Figure 3.4.3 illustrates the rollout algorithm, which is just the first iteration of PI.

In contrast to approximation in value space, the interpretation of PI in terms of Newton's method has a long history. We refer to the original works for linear quadratic problems by Kleinman [Klei68],† and for finite-state infinite horizon discounted and Markov game problems by Pollatschek and Avi-Itzhak [PoA69] (who also showed that the method may oscillate in the game case). Subsequent works, which discuss algorithmic variations and approximations, include Hewer [Hew71], Puterman and Brumelle [PuB78], [PuB79], Santos and Rust [SaR04], Bokanowski, Maroso, and Zidani [BMZ09], Hylla [Hyl11], Magirou, Vassalos, and Barakitis [MVB20], Bertsekas [Ber21c], and Kundu and Kunitsch [KuK21]. Some of these papers address broader classes of problems (such as continuous-time optimal control, minimax problems, and Markov games), and include superlinear convergence rate results under various (often restrictive) assumptions, as well as PI variants. Early related works for control system design include Saridis and Lee [SaL79], Beard [Bea95], and Beard, Saridis, and Wen [BSW99].

---

† This was part of Kleinman's Ph.D. thesis [Kle67] at M.I.T., supervised by M. Athans. Kleinman gives credit for the one-dimensional version of his results to Bellman and Kalaba [BeK65]. Note also that the first proposal of the PI method was given by Bellman in his classic book [Bel57], under the name "approximation in policy space."

Sec. 3.4   Policy Iteration, Rollout, and Newton's Method   53

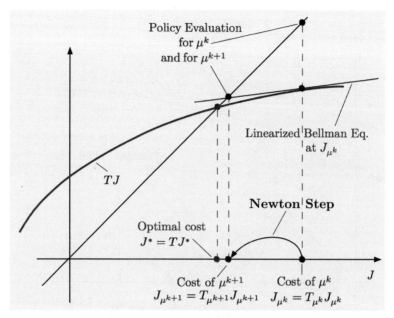

Figure 3.4.2 Geometric interpretation of a policy iteration. Starting from the stable current policy μ^k, it evaluates the corresponding cost function J_{μ^k}, and computes the next policy μ^{k+1} according to

$$T_{\mu^{k+1}} J_{\mu^k} = T J_{\mu^k}.$$

The corresponding cost function $J_{\mu^{k+1}}$ is obtained as the solution of the linearized equation $J = T_{\mu^{k+1}} J$, so it is the result of a Newton step for solving the Bellman equation $J = TJ$, starting from J_{μ^k}. Note that in policy iteration, the Newton step always starts at a function J_μ, which satisfies $J_\mu \geq J^*$ as well as $TJ_\mu \leq J_\mu$ (cf. our discussion on stability in Section 3.3).

Rollout

Generally, rollout with a stable base policy μ can be viewed as a single iteration of Newton's method starting from J_μ, as applied to the solution of the Bellman equation (see Fig. 3.4.3). Note that rollout/policy improvement is applied just at the current state during real-time operation of the system. This makes the on-line implementation possible, even for problems with very large state space, provided that the policy evaluation of the base policy can be done on-line as needed. For this we often need on-line deterministic or stochastic simulation from each of the states x_k generated by the system in real time.

As Fig. 3.4.3 illustrates, the cost function of the rollout policy $J_{\tilde\mu}$ is obtained by constructing a linearized version of Bellman's equation at J_μ (its linear approximation at J_μ), and then solving it. If the function TJ is nearly linear (i.e., has small "curvature") the rollout policy performance

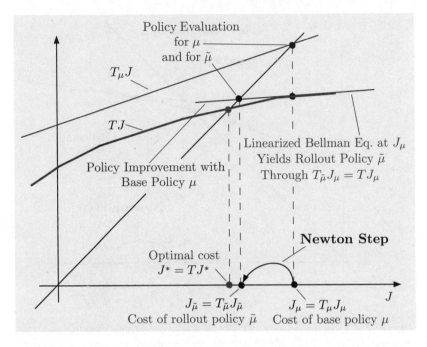

Figure 3.4.3 Geometric interpretation of rollout. Each policy μ defines the linear function $T_\mu J$ of J, given by Eq. (3.2), and TJ is the function given by Eq. (3.1), which can also be written as $TJ = \min_\mu T_\mu J$. The figure shows a policy iteration starting from a base policy μ. It computes J_μ by policy evaluation (by solving the linear equation $J = T_\mu J$ as shown). It then performs a policy improvement using μ as the base policy to produce the rollout policy $\tilde\mu$ as shown: the cost function of the rollout policy, $J_{\tilde\mu}$, is obtained by solving the version of Bellman's equation that is linearized at the point J_μ, as in Newton's method.

$J_{\tilde\mu}(x)$ is very close to the optimal $J^*(x)$, even if the base policy μ is far from optimal. This explains the large cost improvements that are typically observed in practice with the rollout algorithm.

An interesting question is how to compare the rollout performance $J_{\tilde\mu}(x)$ for a given initial state x, with the base policy performance $J_\mu(x)$. Clearly, we would like $J_\mu(x) - J_{\tilde\mu}(x)$ to be large, but this is not the right way to look at cost improvement. The reason is that $J_\mu(x) - J_{\tilde\mu}(x)$ will be small if its upper bound, $J_\mu(x) - J^*(x)$, is small, i.e., if the base policy is close to optimal. What is important is that the error ratio

$$\frac{J_{\tilde\mu}(x) - J^*(x)}{J_\mu(x) - J^*(x)} \qquad (3.17)$$

is small. Indeed, this ratio becomes smaller as $J_\mu(x) - J^*(x)$ approaches 0 because of the superlinear convergence rate of Newton's method that underlies the rollout algorithm (cf. Fig. 3.4.3). Unfortunately, it is hard to evaluate this ratio, since we do not know $J^*(x)$. On the other hand, we

Sec. 3.4   Policy Iteration, Rollout, and Newton's Method   55

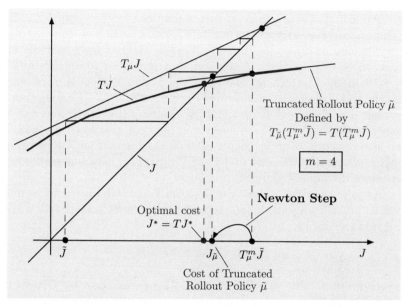

Figure 3.4.4 Geometric interpretation of truncated rollout with one-step lookahead minimization, m value iterations with the base policy μ, and a terminal cost function approximation $\tilde{J}$ (here $m = 4$).

should not be underwhelmed if we observe a small performance improvement $J_\mu(x) - J_{\tilde{\mu}}(x)$: the reason may be that the base policy is already near-optimal, and in fact we are likely doing very well in terms of the ratio (3.17).

Truncated Rollout and Optimistic Policy Iteration

Variants of rollout may involve multistep lookahead, truncation, and terminal cost function approximation, as in the case of AlphaZero/TD-Gammon, cf. Chapter 1. These variants admit geometric interpretations that are similar to the ones given earlier; see Fig. 3.4.4. Truncated rollout uses m VIs with the base policy μ and a terminal cost function approximation $\tilde{J}$ to approximate the cost function J_μ.

In the case of one-step lookahead, the truncated rollout policy $\tilde{\mu}$ is defined by

$$T_{\tilde{\mu}}(T_\mu^m \tilde{J}) = T(T_\mu^m \tilde{J}), \tag{3.18}$$

i.e., $\tilde{\mu}$ attains the minimum when the Bellman operator T is applied to the function $T_\mu^m \tilde{J}$ (the cost obtained by using the base policy μ for m steps followed by terminal cost approximation $\tilde{J}$); see Fig. 3.4.4. In the case of ℓ-step lookahead, the truncated rollout policy $\tilde{\mu}$ is defined by

$$T_{\tilde{\mu}}(T^{\ell-1} T_\mu^m \tilde{J}) = T(T^{\ell-1} T_\mu^m \tilde{J}). \tag{3.19}$$

Truncated rollout is related to a variant of PI called *optimistic*. This variant approximates the policy evaluation step by using m value iterations using the base policy μ; see [BeT96], [Ber12], [Ber19a] for a more detailed discussion of this relation. A method that is related to optimistic PI is the λ-PI method, which is related to the proximal algorithm of convex analysis, and is discussed in several of the author's books ([BeT96], [Ber12], [Ber20a], [Ber22a]), and papers (BeI96], [Ber15], [Ber18d]), and can also be used to define the one-step lookahead policy in place of Eq. (3.18). In particular, Section 6 of the paper [Ber18d] is focused on λ-PI methods, which serve as approximations to the ordinary PI/Newton methods for finite-state discounted and SSP problems.

As noted earlier, variants of Newton's method that involve multiple fixed point iterations, before and after each Newton step, but without truncated rollout, i.e.,

$$T_{\tilde{\mu}}(T^{\ell-1}\tilde{J}) = T(T^{\ell-1}\tilde{J}), \tag{3.20}$$

are well-known. The classical numerical analysis book by Ortega and Rheinboldt [OrR70] (Sections 13.3 and 13.4) provides various convergence results, under assumptions that include differentiability and convexity of the components of T, and nonnegativity of the inverse Jacobian of T. These assumptions, particularly differentiability, may not be satisfied within our DP context. Moreover, for methods of the form (3.20), the initial point must satisfy an additional assumption, which ensures that the convergence to J^* is monotonic from above (in this case, if in addition the Jacobian of T is isotone, an auxiliary sequence can be constructed that converges to J^* monotonically from below; see [OrR70], 13.3.4, 13.4.2). This is similar to existing convergence results for the optimistic PI method in DP; see e.g., [BeT96], [Ber12].

Geometrical interpretations such as the ones of Fig. 3.4.4 suggest, among others, that:

(a) The cost improvement $J_\mu - J_{\tilde{\mu}}$, from base to rollout policy, tends to become larger as the length ℓ of the lookahead increases.

(b) Truncated rollout with ℓ-step lookahead minimization, followed by m steps of a base policy μ, and then followed by terminal cost function approximation $\tilde{J}$ may be viewed, under certain conditions, as an economic alternative to $(\ell+m)$-step lookahead minimization using $\tilde{J}$ as terminal cost function approximation.

Figure 3.4.5 illustrates in summary the approximation in value space scheme with ℓ-step lookahead minimization and m-step truncated rollout [cf. Eq. (3.19)], and its connection to Newton's method. This figure indicates the parts that are ordinarily associated with on-line play and off-line training, and parallels our earlier Fig. 1.2.1, which applies to AlphaZero, TD-Gammon, and related on-line schemes.

Sec. 3.4   Policy Iteration, Rollout, and Newton's Method   57

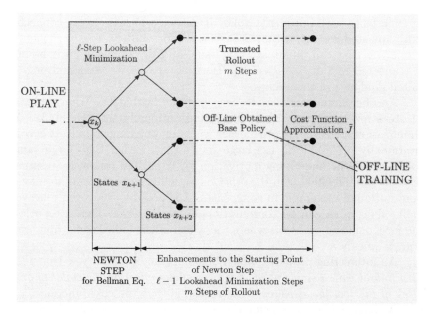

Figure 3.4.5 Illustration of the approximation in value space scheme with ℓ-step lookahead minimization ($\ell = 2$ in the figure) and m-step truncated rollout [cf. Eq. (3.19)], and its connection to Newton's method. The Newton step for solving the Bellman equation $J^* = TJ^*$ corresponds to the first (out of ℓ steps of) lookahead minimization. The remaining $\ell - 1$ steps of lookahead minimization (VI iterations), and the m truncated rollout steps (VI iterations with the base policy), improve the starting point of the Newton step, from its off-line-obtained cost function approximation $\tilde{J}$.

Lookahead Length Issues in Truncated Rollout

A question of practical interest is how to choose the lookahead lengths ℓ and m in truncated rollout schemes. It is clear that large values ℓ for lookahead minimization are beneficial (in the sense of producing improved lookahead policy cost functions $J_{\tilde{\mu}}$), since additional VI iterations bring closer to J^* the starting point $T^{\ell-1}\tilde{J}$ of the Newton step. Note, however, that while long lookahead minimization is computationally expensive (its complexity increases exponentially with ℓ), it is only the first stage of the multistep lookahead that contributes to the Newton step, while the remaining $\ell - 1$ steps are far less effective first order/VI iterations.

Regarding the value of m, long truncated rollout brings the starting point for the Newton step closer to J_μ, but not necessarily closer to J^*, as indicated by Fig. 3.4.4. Indeed computational experiments suggest that increasing values for m may be counterproductive beyond some threshold, and that this threshold generally depends on the problem and the terminal cost approximation $\tilde{J}$; see also our subsequent discussion for linear quadratic problems in Section 4.6. This is also consistent with long standing experience with optimistic policy iteration, which is closely connected with truncated rollout, as noted earlier. Unfortunately, however, there is no analysis that can illuminate this issue, and the available error bounds for truncated rollout (see [Ber19a], [Ber20a]) are conservative and provide limited guidance in this regard.

Another important fact to keep in mind is that the truncated rollout steps are much less demanding computationally than the lookahead minimization steps. Thus, with other concerns weighing equally, it is computationally preferable to use large values of m rather than large values of ℓ (this was the underlying motivation for truncated rollout in Tesauro's TD-Gammon [TeG96]). On the other hand, while large values of m may be computationally tolerable in some cases, it is possible that even relatively small values of m can be computationally prohibitive. This is especially true for stochastic problems where the width of the lookahead graph tends to grow quickly.

An interesting property, which holds in some generality, is that *truncated rollout with a stable policy has a beneficial effect on the stability properties of the lookahead policy*. The reason is that the cost function J_μ of the base policy μ lies well inside the region of stability, as noted in Section 3.2. Moreover value iterations with μ (i.e., truncated rollout) tend to push the starting point of the Newton step towards J_μ. Thus a sufficient number of these value iterations will bring the starting point of the Newton step within the region of stability.

The preceding discussion suggests the following qualitative question: *is lookahead by rollout an economic substitute for lookahead by minimization?* The answer to this seems to be a qualified yes: for a given computational budget, judiciously balancing the values of m and ℓ tends to give better lookahead policy performance than simply increasing ℓ as much as possible, while setting $m = 0$ (which corresponds to no rollout). This is consistent with intuition obtained through geometric constructions such as Fig. 3.4.4, but it is difficult to establish conclusively. We discuss this issue further in Section 4.6 for the case of linear quadratic problems.

3.5 HOW SENSITIVE IS ON-LINE PLAY TO THE OFF-LINE TRAINING PROCESS?

An important issue to consider in approximation in value space is errors in the one-step or multistep minimization, or in the choice of terminal cost approximation $\tilde{J}$. Such errors are often unavoidable because the control constraint set $U(x)$ is infinite, or because the minimization is simplified for reasons of computational expediency (see our subsequent discussion of multiagent problems). Moreover, to these errors, we may add the effect of errors due to rollout truncation, and errors due to changes in problem parameters, which are reflected in changes in Bellman's equation (see our subsequent discussion of robust and adaptive control).

Sec. 3.5 How Sensitive is On-Line Play?

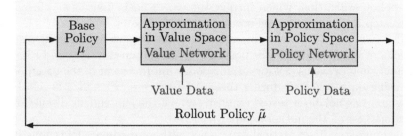

Figure 3.5.1 Schematic illustration of approximate PI. Either the policy evaluation and policy improvement phases (or both) are approximated with a value or a policy network, respectively. These could be neural networks, which are trained with (state, cost function value) data that is generated using the current base policy μ, and with (state, rollout policy control) data that is generated using the rollout policy $\tilde{\mu}$.

Note that there are three different types of approximate implementation involving: 1) a value network but no policy network (here the value network defines a policy via one-step or multistep lookahead), or 2) a policy network but no value network (here the policy network has a corresponding value function that can be computed by rollout), or 3) both a policy and a value network (the approximation architecture of AlphaZero is a case in point).

Under these circumstances the linearization of the Bellman equation at the point $\tilde{J}$ in Fig. 3.4.4 is perturbed, and the corresponding point $T_\mu^m \tilde{J}$ in Fig. 3.4.4 is also perturbed. However, the effect of these perturbations tends to be mitigated by the Newton step that produces the policy $\tilde{\mu}$ and the corresponding cost function $J_{\tilde{\mu}}$. The Newton step has a superlinear convergence property, so for an $O(\epsilon)$-order error [i.e., $O(\epsilon)/\epsilon$ stays bounded as $\epsilon \to 0$] in the calculation of $T_\mu^m \tilde{J}$, the error in $J_{\tilde{\mu}}$ will be of the much smaller order $o(\epsilon)$ [i.e., $o(\epsilon)/\epsilon \to 0$ as $\epsilon \to 0$], when $J_{\tilde{\mu}}$ is near J^*.† This is a significant insight, as it suggests that *extreme accuracy and fine-tuning of the choice of $\tilde{J}$ may not produce significant effects in the resulting performance of the one-step and particularly a multistep lookahead policy*; see also the quantitative analysis for linear quadratic problems in Section 4.5.

Approximate Policy Iteration and Implementation Errors

Both policy evaluation and policy improvement can be approximated, possibly by using training with data and approximation architectures, such as neural networks; see Fig. 3.5.1. Other approximations include simulation-based methods such as truncated rollout, and temporal difference methods

---

† A rigorous proof of this requires differentiability of T at $\tilde{J}$. Since T is differentiable at almost all points J, the sensitivity property just stated, will likely hold in practice even if T is not differentiable. See the appendix, which also compares global and local error bounds for approximation in value space, and for approximate PI (cf. Sections A.3 and A.4)

for policy evaluation, which involve the use of basis functions. Moreover, multistep lookahead may be used in place of one-step lookahead, and simplified minimization, based for example on multiagent rollout, may also be used. Let us also mention the possibility of a combined rollout and PI algorithm, whereby we use PI for on-line policy improvement of the base policy, by using data collected during the rollout process. This idea is relatively new and has not been tested extensively; see the subsequent discussion in Section 3.8 and the author's paper [Ber21a].

Long-standing practical experience with approximate PI is consistent with the view of the effect of implementation errors outlined above, and suggests that substantial changes in the policy evaluation and policy improvement operations often have small but largely unpredictable effects on the performance of the policies generated. For example, when TD(λ)-type methods are used for policy evaluation, the choice of λ has a large effect on the corresponding policy cost function approximations, but often has little and unpredictable effect on the performance of the generated policies through on-line play. A plausible conjecture here is that the superlinear convergence property of the exact Newton step "smooths out" the effect of off-line approximation errors.

3.6 WHY NOT JUST TRAIN A POLICY NETWORK AND USE IT WITHOUT ON-LINE PLAY?

This is a sensible and common question, which stems from the mindset that neural networks have extraordinary function approximation properties. In other words, why go through the arduous on-line process of lookahead minimization, if we can do the same thing off-line and represent the lookahead policy with a trained policy network? More generally, it is possible to use *approximation in policy space*, a major alternative approach to approximation in value space, whereby we select the policy from a suitably restricted class of policies, such as a parametric class of the form $\mu(x, r)$, where r is a parameter vector. We may then estimate r using some type of off-line training process. There are quite a few methods for performing this type of training, such as policy gradient and random search methods (see the books [SuB18] and [Ber19a] for an overview). Alternatively, some approximate DP or classical control system design method may be used.

An important advantage of approximation in policy space is that once the parametrized policy is obtained, the on-line computation of controls $\mu(x, r)$ is often much faster compared with on-line lookahead minimization. For this reason, approximation in policy space can be used to provide an approximate implementation of a known policy (no matter how obtained) for the purpose of convenient use. On the negative side, because parametrized approximations often involve substantial calculations, they are not well suited for on-line replanning.

Sec. 3.7 Multiagent Problems and Multiagent Rollout

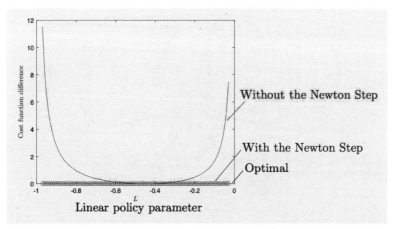

Figure 3.6.1 Illustration of the performance enhancement obtained by rollout with an off-line trained base policy for the linear quadratic problem. Here the system equation is $x_{k+1} = x_k + 2u_k$, and the cost function parameters are $q = 1$, $r = 0.5$. The optimal policy is $\mu^*(x) = L^*x$ with $L^* \approx -0.4$, and the optimal cost function is $J^*(x) = K^*x^2$, where $K^* \approx 1.1$. We consider policies of the form $\mu(x) = Lx$, where L is the parameter, with cost function of the form $J_\mu(x) = K_L x^2$. The figure shows the quadratic cost coefficient differences $K_L - K^*$ and $K_{\tilde{L}} - K^*$ as a function of L, where K_L and $K_{\tilde{L}}$ are the quadratic cost coefficients of μ (without one-step lookahead/Newton step) and the corresponding one-step lookahead policy $\tilde{\mu}$ (with one-step lookahead/Newton step).

From our point of view in this book, there is another important reason why approximation in value space is needed on top of approximation in policy space: *the off-line trained policy may not perform nearly as well as the corresponding one-step or multistep lookahead/rollout policy*, because it lacks the extra power of the associated exact Newton step (cf. our discussion of AlphaZero and TD-Gammon in Chapter 1). Figure 3.6.1 illustrates this fact with a one-dimensional linear-quadratic example, and compares the performance of a linear policy, defined by a scalar parameter, with its corresponding one-step lookahead policy.

3.7 MULTIAGENT PROBLEMS AND MULTIAGENT ROLLOUT

A major difficulty in the implementation of value space approximation with one-step lookahead is the minimization operation over $U(x)$ at a state x. When $U(x)$ is infinite, or even when it is finite but has a very large number of elements, the minimization may become prohibitively time consuming. In the case of multistep lookahead the computational difficulty becomes even more acute. In this section we discuss how to deal with this difficulty when the control u consists of m components, $u = (u_1, \ldots, u_m)$, with a separable control constraint for each component, $u_\ell \in U_\ell(x)$, $\ell = 1, \ldots, m$.

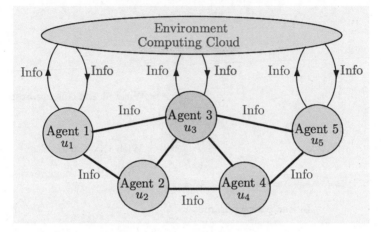

Figure 3.7.1 Schematic illustration of a multiagent problem. There are multiple "agents," and each agent $\ell = 1, \ldots, m$, controls its own decision variable u_ℓ. At each stage, agents exchange new information and also exchange information with the "environment," and then select their decision variables for the stage.

Thus the control constraint set is the Cartesian product

$$U(x) = U_1(x) \times \cdots \times U_m(x), \tag{3.21}$$

where the sets $U_\ell(x)$ are given. This structure is inspired by applications involving distributed decision making by multiple agents with communication and coordination between the agents; see Fig. 3.7.1.

To illustrate our approach, let us consider the discounted infinite horizon problem, and for the sake of the following discussion, assume that each set $U_\ell(x)$ is finite. Then the one-step lookahead minimization of the standard rollout scheme with base policy μ is given by

$$\tilde{u} \in \arg\min_{u \in U(x)} E\Big\{g(x,u,w) + \alpha J_\mu\big(f(x,u,w)\big)\Big\}, \tag{3.22}$$

and involves as many as n^m terms, where n is the maximum number of elements of the sets $U_\ell(x)$ [so that n^m is an upper bound to the number of controls in $U(x)$, in view of its Cartesian product structure (3.21)]. Thus the standard rollout algorithm requires an exponential [order $O(n^m)$] number of computations per stage, which can be overwhelming even for moderate values of m.

This potentially large computational overhead motivates a far more computationally efficient rollout algorithm, whereby the one-step lookahead minimization (3.22) is replaced by a sequence of m successive minimizations, *one-agent-at-a-time*, with the results incorporated into the subsequent minimizations. In particular, at state x we perform the sequence of

Sec. 3.7   Multiagent Problems and Multiagent Rollout

minimizations

$$\tilde{\mu}_1(x) \in \arg\min_{u_1 \in U_1(x)} E_w \Big\{ g\big(x, u_1, \mu_2(x), \ldots, \mu_m(x), w\big)$$
$$+ \alpha J_\mu\big(f(x, u_1, \mu_2(x), \ldots, \mu_m(x), w)\big) \Big\},$$

$$\tilde{\mu}_2(x) \in \arg\min_{u_2 \in U_2(x)} E_w \Big\{ g\big(x, \tilde{\mu}_1(x), u_2, \mu_3(x) \ldots, \mu_m(x), w\big)$$
$$+ \alpha J_\mu\big(f(x, \tilde{\mu}_1(x), u_2, \mu_3(x), \ldots, \mu_m(x), w)\big) \Big\},$$

$$\cdots \quad \cdots \quad \cdots \quad \cdots$$

$$\tilde{\mu}_m(x) \in \arg\min_{u_m \in U_m(x)} E_w \Big\{ g\big(x, \tilde{\mu}_1(x), \tilde{\mu}_2(x), \ldots, \tilde{\mu}_{m-1}(x), u_m, w\big)$$
$$+ \alpha J_\mu\big(f(x, \tilde{\mu}_1(x), \tilde{\mu}_2(x), \ldots, \tilde{\mu}_{m-1}(x), u_m, w)\big) \Big\}.$$

Thus each agent component u_ℓ is obtained by a minimization with the preceding agent components $u_1, \ldots, u_{\ell-1}$ fixed at the previously computed values of the rollout policy, and the following agent components $u_{\ell+1}, \ldots, u_m$ fixed at the values given by the base policy. This algorithm requires order $O(nm)$ computations per stage, a potentially huge computational saving over the order $O(n^m)$ computations required by standard rollout.

A key idea here is that the computational requirements of the rollout one-step minimization (3.22) are proportional to the number of controls in the set $U(x)$ and are independent of the size of the state space. This motivates a reformulation of the problem, first suggested in the book [BeT96], Section 6.1.4, whereby control space complexity is traded off with state space complexity, by "unfolding" the control u_k into its m components, which are applied *one-agent-at-a-time* rather than all-agents-at-once.

In particular, we can reformulate the problem by breaking down the collective decision u_k into m sequential component decisions, thereby reducing the complexity of the control space while increasing the complexity of the state space. The potential advantage is that the extra state space complexity does not affect the computational requirements of some RL algorithms, including rollout.

To this end, we introduce a modified but equivalent problem, involving one-at-a-time agent control selection. At the generic state x, we break down the control u into the sequence of the m controls $u_1, u_2, \ldots, u_m$, and between x and the next state $\bar{x} = f(x, u, w)$, we introduce artificial intermediate "states" $(x, u_1), (x, u_1, u_2), \ldots, (x, u_1, \ldots, u_{m-1})$, and corresponding transitions. The choice of the last control component u_m at "state" $(x, u_1, \ldots, u_{m-1})$ marks the transition to the next state $\bar{x} = f(x, u, w)$ according to the system equation, while incurring cost $g(x, u, w)$; see Fig. 3.7.2.

It is evident that this reformulated problem is equivalent to the original, since any control choice that is possible in one problem is also possible

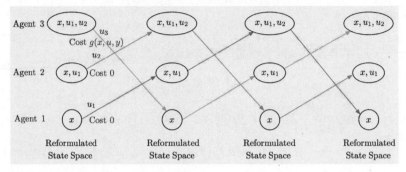

Figure 3.7.2 Equivalent formulation of the stochastic optimal control problem for the case where the control u consists of m components $u_1, u_2, \ldots, u_m$:

$$u = (u_1, \ldots, u_m) \in U_1(x_k) \times \cdots \times U_m(x_k).$$

The figure depicts the kth stage transitions. Starting from state x, we generate the intermediate states

$$(x, u_1), (x_k, u_1, u_2), \ldots, (x, u_1, \ldots, u_{m-1}),$$

using the respective controls $u_1, \ldots, u_{m-1}$. The final control u_m leads from $(x, u_1, \ldots, u_{m-1})$ to $\bar{x} = f(x, u, w)$, and the random cost $g(x, u, w)$ is incurred.

in the other problem, while the cost structure of the two problems is the same. In particular, every policy $\big(\mu_1(x), \ldots, \mu_m(x)\big)$ of the original problem, is admissible for the reformulated problem, and has the same cost function for the original as well as the reformulated problem. Reversely, every policy for the reformulated problem can be converted into a policy for the original problem that produces the same state and control trajectories and has the same cost function.

The motivation for the reformulated problem is that the control space is simplified at the expense of introducing $m-1$ additional layers of states, and the corresponding $m-1$ cost functions

$$J^1(x, u_1), J^2(x, u_1, u_2), \ldots, J^{m-1}(x, u_1, \ldots, u_{m-1}).$$

The increase in size of the state space does not adversely affect the operation of rollout, since the minimization (3.22) is performed for just one state at each stage.

A major fact that follows from the preceding reformulation is that despite the dramatic reduction in computational cost, *multiagent rollout still achieves cost improvement*:

$$J_{\tilde{\mu}}(x) \leq J_\mu(x), \qquad \text{for all } x,$$

where $J_\mu(x)$ is the cost function of the base policy $\mu = (\mu_1, \ldots, \mu_m)$, and $J_{\tilde{\mu}}(x)$ is the cost function of the rollout policy $\tilde{\mu} = (\tilde{\mu}_1, \ldots, \tilde{\mu}_m)$, starting from state x. Furthermore, this cost improvement property can be extended to multiagent PI schemes that involve one-agent-at-a-time policy improvement operations, and have sound convergence properties (see the book [Ber20a], Chapters 3 and 5, as well as the author's papers [Ber19b], [Ber19c], [Ber20b], [Ber21b], and the paper by Bhatacharya et al. [BKB20]).

Another fact that follows from the preceding reformulation is that *multiagent rollout may be viewed as a Newton step applied to the Bellman equation that corresponds to the reformulated problem*, with starting point J_μ. This is very important for our purposes in the context of this book. In particular, *the superlinear cost improvement of the Newton step can still be obtained through multiagent rollout*, even though the amount of computation for the lookahead minimization is dramatically reduced through one-agent-at-a-time minimization. This explains experimental results given in the paper [BKB20], which have shown comparable performance for multiagent and standard rollout in the context of a large-scale multi-robot POMDP application.

Let us also mention that multiagent rollout can become the starting point for various related PI schemes that are well suited for distributed operation in important practical contexts involving multiple autonomous decision makers (see the book [Ber20a], Section 5.3.4, and the papers [Ber21b]).

Multiagent Approximation in Value Space

Let us now consider the reformulated problem of Fig. 3.7.2 and see how we can apply approximation in value space with one-step lookahead minimization, truncated rollout with a base policy $\mu = (\mu_1, \ldots, \mu_m)$, and terminal cost function approximation $\tilde{J}$.

In one such scheme that involves one-agent-at-a-time minimization, at state x we perform the sequence of minimizations

$$\tilde{u}_1 \in \arg\min_{u_1 \in U_1(x)} E_w \Big\{ g\big(x, u_1, \mu_2(x), \ldots, \mu_m(x), w\big)$$
$$+ \alpha \tilde{J}\big(f(x, u_1, \mu_2(x), \ldots, \mu_m(x), w)\big) \Big\},$$

$$\tilde{u}_2 \in \arg\min_{u_2 \in U_2(x)} E_w \Big\{ g\big(x, \tilde{u}_1, u_2, \mu_3(x) \ldots, \mu_m(x), w\big)$$
$$+ \alpha \tilde{J}\big(f(x, \tilde{u}_1, u_2, \mu_3(x), \ldots, \mu_m(x), w)\big) \Big\},$$

$$\ldots \quad \ldots \quad \ldots \quad \ldots$$

$$\tilde{u}_m \in \arg\min_{u_m \in U_m(x)} E_w \Big\{ g\big(x, \tilde{u}_1, \tilde{u}_2, \ldots, \tilde{u}_{m-1}, u_m, w\big)$$
$$+ \alpha \tilde{J}\big(f(x, \tilde{u}_1, \tilde{u}_2, \ldots, \tilde{u}_{m-1}, u_m, w)\big) \Big\}.$$

In the context of the reformulated problem, this is a sequence of one-step lookahead minimizations at the states $x, (x, \tilde{u}_0), \ldots, (x, \tilde{u}_0, \ldots, \tilde{u}_{m-1})$ of the reformulated problem, using truncated rollout with base policy μ of corresponding length $m-1, m-2, \ldots, 0$. The Newton step interpretation of Section 3.4 and Fig. 3.4.4 still applies, with its superlinear convergence rate. At the same time, the computational requirements are dramatically reduced, similar to the multiagent rollout method discussed earlier.

Let us finally note that there are variations of the multiagent schemes of this section, which involve multistep lookahead minimization as well as truncated rollout. They likely result in better performance at the expense of greater computational cost.

3.8 ON-LINE SIMPLIFIED POLICY ITERATION

In this section, we describe some variants of the PI algorithm, introduced in the author's recent paper [Ber21a], which are consistent with the approximation in value space theme of this work. The salient feature of these variants is that *they involve an exact Newton step and are suitable for on-line implementation*, while still maintaining the principal character of PI, which we have viewed so far as an off-line algorithm.

Thus the algorithms of this section involve training and self-improvement through on-line experience. They are also simplified relative to standard PI in two ways:

(a) They perform policy improvement operations only for the states that are encountered during the on-line operation of the system.

(b) The policy improvement operation is simplified in that it uses approximate minimization in the Bellman equation of the current policy at the current state.

Despite these simplifications, we show that our algorithms generate a sequence of improved policies, which converge to a policy with a local optimality property. Moreover, with an enhancement of the policy improvement operation, which involves a form of exploration, they converge to a globally optimal policy.

The motivation comes from the rollout algorithm, which starts from some available base policy and implements on-line an improved rollout policy. In the algorithm of the present section, the data accumulated from the rollout implementation are used to improve on-line the base policy, and to asymptotically obtain a policy that is either locally or globally optimal.

We focus on a finite-state discounted Markov decision problem, with states $1, \ldots, n$, and we use a transition probability notation. We denote states by the symbol x and successor states by the symbol y. The control/action is denoted by u, and is constrained to take values in a given finite constraint set $U(x)$, which may depend on the current state x. The use of a control u at state x specifies the transition probability $p_{xy}(u)$ to the next state y, at a cost $g(x, u, y)$.

Sec. 3.8 On-Line Simplified Policy Iteration

The cost of a policy μ starting from state x_0 is given by

$$J_\mu(x_0) = \lim_{N\to\infty} E\left\{ \sum_{k=0}^{N-1} \alpha^k g\big(x_k, \mu(x_k), x_{k+1}\big) \,\Big|\, x_0, \mu \right\}, \qquad x_0 = 1,\ldots,n,$$

where $\alpha < 1$ is the discount factor. As earlier, J_μ is viewed as the vector in the n-dimensional Euclidean space $\Re^n$, with components $J_\mu(1), \ldots, J_\mu(n)$.

In terms of our abstract notation, for each policy μ, the Bellman operator T_μ maps a vector $J \in \Re^n$ to the vector $T_\mu J \in \Re^n$ that has components

$$(T_\mu J)(x) = \sum_{y=1}^n p_{xy}\big(\mu(x)\big) \Big(g\big(x, \mu(x), y\big) + \alpha J(y)\Big), \qquad x = 1,\ldots,n. \tag{3.23}$$

The Bellman operator $T : \Re^n \mapsto \Re^n$ is given by

$$(TJ)(x) = \min_{u \in U(x)} \sum_{y=1}^n p_{xy}(u)\big(g(x,u,y) + \alpha J(y)\big), \qquad x = 1,\ldots,n. \tag{3.24}$$

For the discounted problem that we consider, the operators T_μ and T are sup-norm contractions (generally this is true for discounted problems with bounded cost per stage [Ber22a]; in our context, the number of states is finite, so the cost per stage is bounded). Thus J_μ is the unique solution of Bellman's equation $J = T_\mu J$, or equivalently

$$J_\mu(x) = \sum_{y=1}^n p_{xy}\big(\mu(x)\big)\Big(g\big(x, \mu(x), y\big) + \alpha J_\mu(y)\Big), \qquad x = 1,\ldots,n. \tag{3.25}$$

Moreover, J^* is the unique solution of Bellman's equation $J = TJ$, so that

$$J^*(x) = \min_{u \in U(x)} \sum_{y=1}^n p_{xy}(u)\big(g(x,u,y) + \alpha J^*(y)\big), \qquad x = 1,\ldots,n. \tag{3.26}$$

Furthermore, the following optimality conditions hold

$$T_\mu J^* = T J^* \qquad \text{if and only if} \qquad \mu \text{ is optimal}, \tag{3.27}$$

$$T_\mu J_\mu = T J_\mu \qquad \text{if and only if} \qquad \mu \text{ is optimal}. \tag{3.28}$$

The contraction property also implies that the VI algorithms

$$J^{k+1} = T_\mu J^k, \qquad J^{k+1} = T J^k$$

generate sequences $\{J^k\}$ that converge to J_μ and J^*, respectively, from any starting vector $J^0 \in \Re^n$.

As discussed earlier, in the PI algorithm, the current policy μ is improved by finding $\tilde{\mu}$ that satisfies

$$T_{\tilde{\mu}} J_\mu = T J_\mu$$

[i.e., by minimizing for all x in the right-hand side of Eq. (3.24) with J_μ in place of J]. The improved policy $\tilde{\mu}$ is evaluated by solving the linear $n \times n$ system of equations $J_{\tilde{\mu}} = T_{\tilde{\mu}} J_{\tilde{\mu}}$, and then $(J_{\tilde{\mu}}, \tilde{\mu})$ becomes the new cost vector-policy pair, which is used to start a new iteration. Thus the PI algorithm starts with a policy μ^0 and generates a sequence $\{\mu^k\}$ according to

$$J_{\mu^k} = T_{\mu^k} J_{\mu^k}, \qquad T_{\mu^{k+1}} J_{\mu^k} = T J_{\mu^k}. \tag{3.29}$$

We now introduce an on-line variant of PI, which starts at time 0 with a state-policy pair (x_0, μ^0) and generates on-line a sequence of state-policy pairs (x_k, μ^k). We view x_k as the current state of a system operating online under the influence of the policies $\mu^1, \mu^2, \ldots$. In our algorithm, μ^{k+1} *may differ from μ^k only at state x_k*. The control $\mu^{k+1}(x_k)$ and the state x_{k+1} are generated as follows:

We consider the right-hand sides of Bellman's equation for μ^k (also known as *Q-factors* of μ^k)

$$Q_{\mu^k}(x_k, u) = \sum_{y=1}^{n} p_{x_k y}(u)\big(g(x_k, u, y) + \alpha J_{\mu^k}(y)\big), \tag{3.30}$$

and we select the control $\mu^{k+1}(x_k)$ from within the constraint set $U(x_k)$ with sufficient accuracy to satisfy the following condition

$$Q_{\mu^k}\big(x_k, \mu^{k+1}(x_k)\big) \le J_{\mu^k}(x_k), \tag{3.31}$$

with strict inequality whenever this is possible.† For $x \ne x_k$ the policy controls are not changed:

$$\mu^{k+1}(x) = \mu^k(x) \text{ for all } x \ne x_k.$$

The next state x_{k+1} is generated randomly according to the transition probabilities $p_{x_k x_{k+1}}\big(\mu^{k+1}(x_k)\big)$.

---

† By this we mean that if $\min_{u \in U(x_k)} Q_{\mu^k}(x_k, u) < J_{\mu^k}(x_k)$ we select a control u_k that satisfies

$$Q_{\mu^k}(x_k, u_k) < J_{\mu^k}(x_k), \tag{3.32}$$

and set $\mu^{k+1}(x_k) = u_k$, and otherwise we set $\mu^{k+1}(x_k) = \mu^k(x_k)$ [so Eq. (3.31) is satisfied]. Such a control selection may be obtained by a number of schemes, including brute force calculation and random search based on Bayesian optimization. The needed values of the Q-factor Q_{μ^k} and cost J_{μ^k} may be obtained in several ways, depending on the problem at hand, including by on-line simulation.

Sec. 3.8 On-Line Simplified Policy Iteration

We first show that the current policy is monotonically improved, i.e., that
$$J_{\mu^{k+1}}(x) \leq J_{\mu^k}(x), \qquad \text{for all } x \text{ and } k,$$
with strict inequality for $x = x_k$ (and possibly other values of x) if
$$\min_{u \in U(x_k)} Q_{\mu^k}(x_k, u) < J_{\mu^k}(x_k).$$

To prove this, we note that the policy update is done under the condition (3.31). By using the monotonicity of $T_{\mu^{k+1}}$, we have for all $\ell \geq 1$,
$$T_{\mu^{k+1}}^{\ell+1} J_{\mu^k} \leq T_{\mu^{k+1}}^{\ell} J_{\mu^k} \leq J_{\mu^k}, \qquad (3.33)$$
so by taking the limit as $\ell \to \infty$ and by using the convergence property of VI ($T_{\mu^{k+1}}^{\ell} J \to J_{\mu^{k+1}}$ for any J), we obtain $J_{\mu^{k+1}} \leq J_{\mu^k}$. Moreover, the algorithm selects $\mu^{k+1}(x_k)$ so that
$$(T_{\mu^{k+1}} J_{\mu^k})(x_k) = Q_{\mu^k}(x_k, u_k) < J_{\mu^k}(x_k)$$
if
$$\min_{u \in U(x_k)} Q_{\mu^k}(x_k, u) < J_{\mu^k}(x_k),$$
[cf. Eq. (3.32)], and then by using Eq. (3.33), we have $J_{\mu^{k+1}}(x_k) < J_{\mu^k}(x_k)$.

Local Optimality

We next discuss the convergence and optimality properties of the algorithm. We introduce a definition of local optimality of a policy, whereby the policy selects controls optimally only within a subset of states.

Given a subset of states S and a policy μ, we say that μ is *locally optimal over S* if μ is optimal for the problem where the control is restricted to take the value $\mu(x)$ at the states $x \notin S$, and is allowed to take any value $u \in U(x)$ at the states $x \in S$.

Roughly speaking, μ is locally optimal over S, if μ is acting optimally within S, but under the (incorrect) assumption that once the state of the system gets to a state x outside S, there will be no option to select control other than $\mu(x)$. Thus if the choices of μ outside of S are poor, its choices within S may also be poor.

Mathematically, μ is locally optimal over S if
$$J_\mu(x) = \min_{u \in U(x)} \sum_{y=1}^n p_{xy}(u)\big(g(x, u, y) + \alpha J_\mu(y)\big), \qquad \text{for all } x \in S,$$
$$J_\mu(x) = \sum_{y=1}^n p_{xy}\big(\mu(x)\big)\big(g(x, \mu(x), y) + \alpha J_\mu(y)\big), \qquad \text{for all } x \notin S,$$

which can be written compactly as

$$(T_\mu J_\mu)(x) = (TJ_\mu)(x), \qquad \text{for all } x \in S. \tag{3.34}$$

Note that this is different than (global) optimality of μ, which holds if and only if the above condition holds for all $x = 1, \ldots, n$, rather than just for $x \in S$ [cf. Eq. (3.28)]. However, it can be seen that a (globally) optimal policy is also locally optimal within any subset of states.

Our main convergence result is the following.

Proposition 3.8.1: Let $\overline{S}$ be the subset of states that are repeated infinitely often within the sequence $\{x_k\}$. Then the corresponding sequence $\{\mu^k\}$ converges finitely to some policy $\overline{\mu}$ in the sense that $\mu^k = \overline{\mu}$ for all k after some index $\overline{k}$. Moreover $\overline{\mu}$ is locally optimal within $\overline{S}$, while $\overline{S}$ is invariant under $\overline{\mu}$, in the sense that

$$p_{xy}(\overline{\mu}(x)) = 0 \qquad \text{for all } x \in \overline{S} \text{ and } y \notin \overline{S}.$$

Proof: The cost function sequence $\{J_{\mu^k}\}$ is monotonically nonincreasing, as shown earlier. Moreover, the number of policies μ is finite in view of the finiteness of the state and control spaces. Therefore, the number of corresponding functions J_μ is also finite, so J_{μ^k} converges in a finite number of steps to some $\bar{J}$, which in view of the algorithm's construction [selecting $u_k = \mu^k(x_k)$ if $\min_{u \in U(x_k)} Q_{\mu^k}(x_k, u) = J_{\mu^k}(x_k)$; cf. Eq. (3.32)], implies that μ^k will remain unchanged at some $\overline{\mu}$ with $J_{\overline{\mu}} = \bar{J}$ after some sufficiently large k.

We will show that the local optimality condition (3.34) holds for $S = \overline{S}$ and $\mu = \overline{\mu}$. In particular, we have $x_k \in \overline{S}$ and $\mu^k = \overline{\mu}$ for all k greater than some index, while for every $x \in \overline{S}$, we have $x_k = x$ for infinitely many k. It follows that for all $x \in \overline{S}$,

$$Q_{\overline{\mu}}(x, \overline{\mu}(x)) = J_{\overline{\mu}}(x), \tag{3.35}$$

while by the construction of the algorithm,

$$Q_{\overline{\mu}}(x, u) \geq J_{\overline{\mu}}(x), \qquad \text{for all } u \in U(x), \tag{3.36}$$

since the reverse would imply that $\mu^{k+1}(x) \neq \mu^k(x)$ for infinitely many k [cf. Eq. (3.32)]. Condition (3.35) can be written as $J_{\overline{\mu}}(x) = (T_{\overline{\mu}} J_{\overline{\mu}})(x)$ for all $x \in \overline{S}$, and combined with Eq. (3.36), implies that

$$(T_{\overline{\mu}} J_{\overline{\mu}})(x) = (T J_{\overline{\mu}})(x), \qquad \text{for all } x \in \overline{S}.$$

Sec. 3.8 On-Line Simplified Policy Iteration

This is the local optimality condition (3.34) with $S = \overline{S}$ and $\mu = \overline{\mu}$.

To show that $\overline{S}$ is invariant under $\overline{\mu}$, we argue by contradiction: if this were not so, there would exist a state $x \in \overline{S}$ and a state $y \notin \overline{S}$ such that $p_{xy}\bigl(\overline{\mu}(x)\bigr) > 0$, implying that y would be generated following the occurrence of x infinitely often within the sequence $\{x_k\}$, and hence would have to belong to $\overline{S}$ (by the definition of $\overline{S}$). **Q.E.D.**

Note an implication of the invariance property of the set $\overline{S}$ shown in the preceding proposition. We have that $\overline{\mu}$ is (globally) optimal under the assumption that for every policy there does not exist any strict subset of states that is invariant.

A Counterexample to Global Optimality

The following deterministic example (given to us by Yuchao Li) shows that the policy $\overline{\mu}$ obtained by the algorithm need not be (globally) optimal. Here there are three states 1, 2, and 3. From state 1 we can go to state 2 at cost 1, and to state 3 at cost 0, from state 2 we can go to states 1 and 3 at cost 0, and from state 3 we can go to state 2 at cost 0 or stay in 3 at a high cost (say 10). The discount factor is $\alpha = 0.9$. Then it can be verified that the optimal policy is

$$\mu^*(1) : \text{Go to 3}, \qquad \mu^*(2) : \text{Go to 3}, \qquad \mu^*(3) : \text{Go to 2},$$

with optimal costs

$$J^*(1) = J^*(2) = J^*(3) = 0,$$

while the policy

$$\overline{\mu}(1) : \text{Go to 2}, \qquad \overline{\mu}(2) : \text{Go to 1}, \qquad \overline{\mu}(3) : \text{Stay at 3},$$

is strictly suboptimal, but is locally optimal over the set of states $\overline{S} = \{1, 2\}$. Moreover our on-line PI algorithm, starting from state 1 and the policy $\mu^0 = \overline{\mu}$, oscillates between the states 1 and 2, and leaves the policy μ^0 unchanged. Note also that $\overline{S}$ is invariant under $\overline{\mu}$, consistent with Prop. 3.8.1.

On-Line Variants of Policy Iteration with Global Optimality Properties

To address the local versus global convergence issue illustrated by the preceding example, we consider an alternative scheme, whereby in addition to u_k, we generate an additional control at a randomly chosen state $\overline{x}_k \neq x_k$.[†] In particular, assume that at each time k, in addition to u_k and x_{k+1} that

---

† It is also possible to choose multiple additional states at time k for a policy improvement operation, and this is well-suited for the use of parallel computation.

are generated according to Eq. (3.32), the algorithm generates randomly another state $\overline{x}_k$ (all states are selected with positive probability), performs a policy improvement operation at that state as well, and modifies accordingly $\mu^{k+1}(\overline{x}_k)$. Thus, in addition to a policy improvement operation at each state within the generated sequence $\{x_k\}$, there is an additional policy improvement operation at each state within the randomly generated sequence $\{\overline{x}_k\}$.

Because of the random mechanism of selecting $\overline{x}_k$, it follows that at every state there will be a policy improvement operation infinitely often, which implies that the policy $\overline{\mu}$ ultimately obtained is (globally) optimal. Note also that *we may view the random generation of the sequence $\{\overline{x}_k\}$ as a form of exploration.* The probabilistic mechanism for generating the random sequence $\{\overline{x}_k\}$ may be guided by some heuristic reasoning, which aims to explore states with high cost improvement potential.

Let us also note the possibility of approximate implementations of the algorithms described above. In particular, one may start with some base policy, which may be periodically updated using some approximation in policy space scheme, while incorporating the policy improvement data generated so far. As long as the most recent policy improvement results are maintained for the states that have been encountered in the past, the convergence results described above will be maintained.

Finally, let us mention that the idea of on-line PI of the present section can be extended to a broader algorithmic context of *on-line improvement of the approximation in value space process*. In particular, we may consider starting the on-line play algorithm with a cost function approximation $\tilde{J}$, obtained through some off-line training process. We may then try to gradually enhance the quality of $\tilde{J}$ through on-line experience. For example, $\tilde{J}$ may be constructed through some form of machine learning or Bayesian optimization method that is capable of improving the approximation using data obtained in the process of on-line play. There are many possibilities along these lines, and they are a fruitful area of research, particularly within the context of specific applications.

3.9 EXCEPTIONAL CASES

Let us now consider situations where exceptional behavior occurs. One such situation is when the Bellman equation $J = TJ$ has multiple solutions. Then the VI algorithm, when started at one of these solutions will stay at that solution. More generally, it may be unclear whether the VI algorithm will converge to J^*, even when started at seemingly favorable initial conditions. Other types of exceptional behavior may also occur, including cases where the Bellman equation has no solution within the set of real-valued functions. The most unusual case of all is when J^* is real-valued but does not satisfy the Bellman equation $J = TJ$, which in turn has other real-valued solutions; see [BeY16] and [Ber22a], Section 3.1. This is a

Sec. 3.9 Exceptional Cases

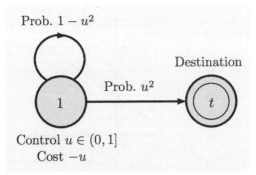

Figure 3.9.1. Transition diagram for the blackmailer problem. At state 1, the blackmailer may demand any amount $u \in (0, 1]$. The victim will comply with probability $1 - u^2$ and will not comply with probability u^2, in which case the process will terminate.

highly unusual phenomenon, which will not be discussed here. It need not be of practical concern, as it arises only in artificial examples; see [BeY16]. Still it illustrates the surprising range of exceptional situations that should be taken into account in theoretical analyses and computational studies.

In this section we provide some examples that illustrate the mechanism by which exceptional behavior in infinite horizon DP can occur, and we highlight the need for rigorous analysis of RL methods when used in contexts that are beyond the well-behaved discounted case, where the Bellman operator is a contraction mapping. For further discussion and analysis that address exceptional behavior, including the frameworks of semicontractive and noncontractive DP, we refer to the author's abstract DP monograph [Ber22a].

The Blackmailer's Dilemma

This is a classical example involving a profit maximizing blackmailer. We formulate it as an SSP problem involving cost minimization, with a single state $x = 1$, in addition to the termination state t. We are in state 1 when the victim is compliant, and we are in state t when the victim refuses to yield to the blackmailer's demand (a refusal is permanent, in the sense that once the blackmailer's demand is refused, all subsequent demands are assumed to be refused, so t is a termination state). At state 1 we can choose a control $u \in (0, 1]$, which we regard as the demand made by the blackmailer. The problem is to find the blackmailer's policy that maximizes his expected total gain.

To formulate this problem as a minimization problem, we will use $(-u)$ as the cost per stage. In particular, upon choosing $u \in (0, 1]$, we move to state t with probability u^2, and stay in state 1 with probability $1 - u^2$; see Fig. 3.9.1. The idea is to optimally balance the blackmailer's

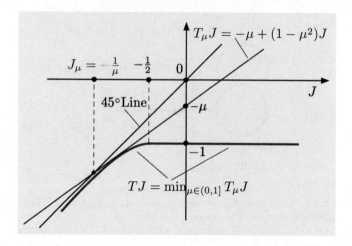

Figure 3.9.2. The Bellman operators and the Bellman equation for the blackmailer problem.

desire for increased demands (large u) with keeping his victim compliant (small u).

For notational simplicity, let us abbreviate $J(1)$ and $\mu(1)$ with just the scalars J and μ, respectively. Then in terms of abstract DP we have $X = \{1\}$, $U = (0,1]$, and for every stationary policy μ, the corresponding Bellman operator T_μ, restricted to state 1 is given by

$$T_\mu J = -\mu + (1-\mu^2)J; \tag{3.37}$$

[at the state t, $J_\mu(t) = 0$]. Clearly T_μ is linear, maps the real line $\Re$ to itself, and is a contraction with modulus $1 - \mu^2$. Its unique fixed point within $\Re$, J_μ, is the solution of

$$J_\mu = T_\mu J_\mu = -\mu + (1-\mu^2)J_\mu,$$

which yields

$$J_\mu = -\frac{1}{\mu};$$

see Fig. 3.9.2. Here all policies are stable and lead asymptotically to t with probability 1, and the infimum of J_μ over $\mu \in (0,1]$ is $-\infty$, implying also that $J^* = -\infty$. However, there is no optimal policy.

The Bellman operator T is given by

$$TJ = \min_{0 < u \leq 1} \{-u + (1-u^2)J\},$$

which after some calculation can be shown to have the form

$$TJ = \begin{cases} -1 & \text{for } -\frac{1}{2} \leq J, \\ J + \frac{1}{4J} & \text{for } J \leq -\frac{1}{2}. \end{cases}$$

Sec. 3.9 Exceptional Cases

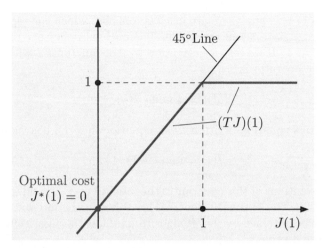

Figure 3.9.3 Illustration of the Bellman equation for a shortest path problem in the exceptional case where there is a cycle of zero length. Restricted within the set of J with $J(t) = 0$, the Bellman operator has the form

$$(TJ)(1) = \min\big\{J(1),\, 1\big\}.$$

The set of solutions of Bellman's equation, $J(1) = (TJ)(1)$ is the interval $(-\infty, 1]$ and contains $J^*(1) = 0$ in its interior.

The form of T is illustrated in Fig. 3.9.2. It can be seen from this figure that the Bellman equation $J = TJ$ has no real-valued solution (the optimal cost $J^* = -\infty$ is a solution within the set of extended real numbers $[-\infty, \infty]$). Moreover the VI algorithm will converge to J^* starting from any $J \in \Re$. It can be verified also that the PI algorithm, starting from any policy $\mu^0 \in (0, 1]$, produces the ever improving sequence of policies $\{\mu^k\}$ with $\mu^{k+1} = \mu^k/2$. Thus μ^k converges to 0, which is not a feasible policy. Also $J_{\mu^k} = -1/\mu^k$, and we have $J_{\mu^k} \downarrow -\infty = J^*$, so the PI algorithm gives in the limit the infinite optimal cost. For additional related examples and discussion relating to the blackmailer problem, see [Ber22a], Section 3.1.

A Shortest Path Problem

Another exceptional type of example is provided by shortest path problems that contain cycles of zero length; see the monograph [Ber22a], Section 3.1. In this case there are infinitely many solutions to Bellman's equation, and the VI and PI algorithms, as well as the approximation in value space process exhibit unusual behavior. We demonstrate this with a shortest path problem involving a single state, denoted 1, in addition to the cost-free destination state t.

In particular, let $X = \{t, 1\}$, and assume that at state 1 there are two options: we can stay at 1 at cost 0, or move to t at cost 1. Here

$J^*(t) = J^*(1) = 0$, and there are just two policies, which correspond to the two options at state 1 and are stable. The optimal policy starting at state 1 is to stay at 1. If we restrict attention to cost functions J with $J(t) = 0$, the Bellman operator is

$$(TJ)(1) = \min\{J(1), 1\},$$

and Bellman's equation, written as an equation in $J(1)$, has the form

$$J(1) = \min\{J(1), 1\}.$$

The set of solutions of this equation is the interval $(-\infty, 1]$ and it is infinite; see Fig. 3.9.3. The optimal value $J^*(1) = 0$ lies in the interior of this set, and cannot be obtained by the VI algorithm, unless the algorithm is started at the optimal value.

Let us consider approximation in value space with cost approximation $\tilde{J}(1)$. Then it can be seen that if $\tilde{J}(1) < 1$, the one-step lookahead policy is to stay at state 1, which is optimal. If $\tilde{J}(1) > 1$, the one-step lookahead policy is to move from state 1 to state t, which is suboptimal. If $\tilde{J}(1) = 1$, either one of the two policies can be the one-step lookahead policy.

Consider also the PI algorithm, starting from the suboptimal policy μ that moves from state 1 to state t. Then $J_\mu(t) = 0$, $J_\mu(1) = 1$, and it can be seen that μ satisfies the policy improvement equation

$$\mu(1) \in \arg\min\{J_\mu(1), 1 + J_\mu(t)\}$$

(the same is true for the optimal policy that stays at state 1). Thus the PI algorithm may stop with the suboptimal policy μ.

Problems where exceptional behavior occurs arise often in Markov decision problems, once one departs from the most commonly discussed and best behaved paradigm of discounted cost with bounded cost per stage, where the mappings T_μ of all policies μ have favorable contraction properties. Moreover, problems arising in decision and control, such as those that have been addressed with MPC, often give rise to exceptional behavior. Further research and computational experimentation is expected to provide improved guidelines for the solution of such problems.

What Happens When the Bellman Operator is Neither Concave nor Convex? - Markov Games

We have discussed so far DP models where the Bellman operator has a concavity property. On the other hand there are interesting DP models where this is not so. An important case in point is *discounted Markov games*, a form of zero-sum games with a dynamic Markov chain structure.

Let us consider two players that play repeated matrix games at each of an infinite number of stages, using mixed strategies. The game played

Sec. 3.9   Exceptional Cases

at a given stage is defined by a state x that takes values in a finite set X, and changes from one stage to the next according to a Markov chain whose transition probabilities are influenced by the players' choices. At each stage and state $x \in X$, the minimizer selects a probability distribution $u = (u_1, \ldots, u_n)$ over n possible choices $i = 1, \ldots, n$, and the maximizer selects a probability distribution $v = (v_1, \ldots, v_m)$ over m possible choices $j = 1, \ldots, m$. If the minimizer chooses i and the maximizer chooses j, the payoff of the stage is $a_{ij}(x)$ and depends on the state x. Thus the expected payoff of the stage is $\sum_{i,j} a_{ij}(x) u_i v_j$ or $u'A(x)v$, where $A(x)$ is the $n \times m$ matrix with components $a_{ij}(x)$ (u and v are viewed as column vectors, and a prime denotes transposition). The two players choose u and v with knowledge of the state x, so they are viewed as using policies μ and ν, where $\mu(x)$ and $\nu(x)$ are the choices of the minimizer and the maximizer, respectively, at a state x.

The state evolves according to transition probabilities $q_{xy}(i,j)$, where i and j are the moves selected by the minimizer and the maximizer, respectively (here y represents the next state and game to be played after moves i and j are chosen at the game represented by x). When the state is x, under u and v, the state transition probabilities are

$$p_{xy}(u,v) = \sum_{i=1}^{n} \sum_{j=1}^{m} u_i v_j q_{xy}(i,j) = u' Q_{xy} v,$$

where Q_{xy} is the $n \times m$ matrix that has components $q_{xy}(i,j)$. Payoffs are discounted by $\alpha \in (0,1)$, and the objectives of the minimizer and maximizer, are to minimize and to maximize the total discounted expected payoff, respectively.

It was shown by Shapley [Sha53] that the problem can be formulated as a fixed point problem involving the mapping H given by

$$\begin{aligned} H(x,u,v,J) &= u'A(x)v + \alpha \sum_{y \in X} p_{xy}(u,v) J(y) \\ &= u' \left(A(x) + \alpha \sum_{y \in X} Q_{xy} J(y) \right) v, \end{aligned} \tag{3.38}$$

with the corresponding Bellman operator given by

$$(TJ)(x) = \min_{u \in U} \max_{v \in V} H(x,u,v,J), \quad \text{for all } x \in X. \tag{3.39}$$

It can be verified that T is an unweighted sup-norm contraction, and its unique fixed point J^* satisfies the Bellman equation $J^* = TJ^*$.

Note that since the matrix defining the mapping H of Eq. (3.38),

$$A(x) + \alpha \sum_{y \in X} Q_{xy} J(y),$$

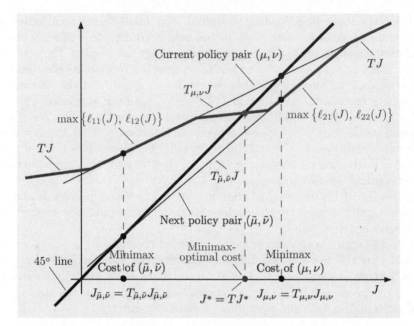

Figure 3.9.4 Schematic illustration of the PI algorithm/Newton's method in the case of a Markov game involving a single state, in addition to a termination state t. We have $J^*(t) = 0$ and $(TJ)(t) = 0$ for all J with $J(t) = 0$, so that the operator T can be graphically represented in just one dimension (denoted by J) that corresponds to the nontermination state. This makes it easy to visualize T and geometrically interpret why Newton's method does not converge. Because the operator T may be neither convex nor concave for a Markov game, the algorithm may cycle between pairs (μ, ν) and $(\tilde{\mu}, \tilde{\nu})$, as shown in the figure. By contrast in a (single-player) finite-state Markov decision problem, $(TJ)(x)$ is piecewise linear and concave, and the PI algorithm converges in a finite number of iterations.

The figure illustrates an operator T of the form

$$TJ = \min \Big\{ \max \big\{ \ell_{11}(J), \ell_{12}(J) \big\}, \max \big\{ \ell_{21}(J), \ell_{22}(J) \big\} \Big\},$$

where $\ell_{ij}(J)$, are linear functions of J, corresponding to the choices $i = 1, 2$ of the minimizer and $j = 1, 2$ of the maximizer. Thus TJ is the minimum of the convex functions

$$\max \big\{ \ell_{11}(J), \ell_{12}(J) \big\} \quad \text{and} \quad \max \big\{ \ell_{21}(J), \ell_{22}(J) \big\},$$

as shown in the figure. Newton's method linearizes TJ at the current iterate [i.e., replaces TJ with one of the four linear functions $\ell_{ij}(J)$, $i = 1, 2$, $j = 1, 2$ (the one attaining the min-max at the current iterate)] and solves the corresponding linear fixed point problem to obtain the next iterate. The figure illustrates a case where the PI algorithm/Newton's method oscillates between two pairs of policies (μ, ν) and $(\tilde{\mu}, \tilde{\nu})$.

is independent of u and v, we may view $J^*(x)$ as the value of a static (nonsequential) matrix game that depends on x. In particular, from a fundamental saddle point theorem for matrix games, we have

$$\min_{u \in U} \max_{v \in V} H(x, u, v, J^*) = \max_{v \in V} \min_{u \in U} H(x, u, v, J^*), \qquad \text{for all } x \in X. \tag{3.40}$$

The paper by Shapley [Sha53] also showed that the strategies obtained by solving the static saddle point problem (3.40) correspond to a saddle point of the sequential game in the space of mixed strategies. Thus once we find J^* as the fixed point of the mapping T [cf. Eq. (3.39)], we can obtain equilibrium policies for the minimizer and maximizer by solving the matrix game (3.40). Moreover, T can be defined via the operator $T_{\mu,\nu}$ defined for a pair of minimizer-maximizer policies (μ, ν) by

$$(T_{\mu,\nu} J)(x) = H\big(x, \mu(x), \nu(x), J\big), \qquad \text{for all } x \in X, \tag{3.41}$$

In particular, T can be defined via a minimax operation applied to the operator $T_{\mu,\nu}$ as follows:

$$(TJ)(x) = \min_{\mu \in \mathcal{M}} \max_{\nu \in \mathcal{N}} (T_{\mu,\nu} J)(x), \qquad \text{for all } x \in X,$$

where $\mathcal{M}$ and $\mathcal{N}$ are the sets of policies of the minimizer and the maximizer, respectively.

On the other hand the Bellman operator components $(TJ)(x)$ may be neither convex nor concave. In particular, the maximization makes the function

$$\max_{v \in V} H(x, u, v, J)$$

convex as a function of x for each fixed $u \in U$, while the subsequent minimization over $u \in U$ tends to introduce concave "pieces" into $(TJ)(x)$. It is possible to apply PI ideas and the corresponding Newton's method to compute the fixed point of T, and in fact this has been proposed by Pollatschek and Avi-Itzhak [PoA69]. However, this algorithm need not converge to the optimal and may not yield J^*, the fixed point of T (unless the starting point is sufficiently close to J^*, as has been recognized in the paper [PoA69]). The mechanism by which this phenomenon may occur is illustrated in Fig. 3.9.4. In fact a two-state example where the PI algorithm/Newton's method does not converge to J^* was given by van der Wal [Van78]. The preceding Markov chain discussion is part of a broader investigation of abstract minimax problems and Markov games, given in the author's recent paper [Ber21c] (and replicated in the monograph [Ber22a], Ch. 5). In particular, this paper develops exact and approximate PI methods, which correct the exceptional behavior illustrated in Fig. 3.9.4.

3.10 NOTES AND SOURCES

The author's abstract DP monograph [Ber22a] (originally published in 2013, with a second edition appearing in 2018, and a third edition appearing in 2022) has provided the framework for the Newton step interpretations and visualizations that we have used to gain insight into approximation in value space, rollout, and policy iteration. The abstract framework aims at a unified development of the core theory and algorithms of total cost sequential decision problems, and addresses simultaneously stochastic, minimax, game, risk-sensitive, and other DP problems, through the use of the abstract DP operator (or Bellman operator as it is often called in RL). The idea here is to gain insight through abstraction. In particular, the structure of a DP model is encoded in its abstract Bellman operator, which serves as the "mathematical signature" of the model. Characteristics of this operator (such as monotonicity and contraction) largely determine the analytical results and computational algorithms that can be applied to that model.

Abstraction also captures the generality of the DP methodology. In particular, our conceptual framework based on Newton's method is applicable to problems with general state and control spaces, ranging from the continuous spaces control problems, traditionally the focus of MPC, to Markov decision problems, traditionally the focus of operations research as well as RL, and to discrete optimization problems, traditionally the focus of integer programming and combinatorial optimization. A key mathematical fact in this respect is that while the state and control spaces may be continuous or discrete, the Bellman operators and equations are always defined over continuous function spaces, and are thus amenable to solution through the use of continuous spaces algorithms, including Newton's method.

4

The Linear Quadratic Case - Illustrations

Contents
4.1. Optimal Solution . . . . p. 82
4.2. Cost Functions of Stable Linear Policies . . . . p. 83
4.3. Value Iteration . . . . p. 86
4.4. One-Step and Multistep Lookahead - Newton Step
Interpretations . . . . p. 86
4.5. Sensitivity Issues . . . . p. 91
4.6. Rollout and Policy Iteration . . . . p. 94
4.7. Truncated Rollout - Length of Lookahead Issues . . . . p. 97
4.8. Exceptional Behavior in Linear Quadratic Problems . . . p. 99
4.9. Notes and Sources . . . . p. 100

In this chapter, we will use linear quadratic problems as a vehicle for graphical illustrations and insight into the suboptimal control ideas developed so far. This is possible because linear quadratic problems admit closed form solutions. Our discussion applies to multidimensional linear quadratic problems (cf. Example 2.1.1), but we will focus on the one-dimensional case to demonstrate graphically our approximation in value space ideas and their connection to Newton's method.

In particular, throughout this chapter we will consider the system

$$x_{k+1} = ax_k + bu_k,$$

and the cost function

$$\sum_{k=0}^{\infty}(qx_k^2 + ru_k^2),$$

where a, b, q, r are scalars with $b \neq 0$, $q > 0$, $r > 0$. It can be verified computationally (and also with some analysis) that the insights obtained from the one-dimensional case are generally correct for the multidimensional case of the linear quadratic problem, where the state cost weighting matrix Q is positive definite. In Section 4.8 we will obtain related insights about what happens in the exceptional case where $q = 0$.

4.1 OPTIMAL SOLUTION

The optimal solution was given for the multidimensional case of the linear quadratic problem in Example 2.1.1. For the one-dimensional case considered here, the optimal cost function has the form

$$J^*(x) = K^*x^2, \qquad (4.1)$$

where the scalar K^* solves a fixed point equation of the form

$$K = F(K), \qquad (4.2)$$

with F defined by

$$F(K) = \frac{a^2 rK}{r + b^2 K} + q. \qquad (4.3)$$

This is the Riccati equation, which is equivalent to the Bellman equation $J = TJ$, restricted to the subspace of quadratic functions of the form $J(x) = Kx^2$; see Fig. 4.1.1. Essentially, *by replacing the Bellman operator T with the Riccati equation operator F of Eq. (4.3), we can analyze the action of T on this subspace.* This allows a different type of visualization than the one we have been using so far.

The scalar K^* that corresponds to the optimal cost function J^* [cf. Eq. (4.1)] is the unique solution of the Riccati equation (4.2) within the

Sec. 4.2 Cost Functions of Stable Linear Policies

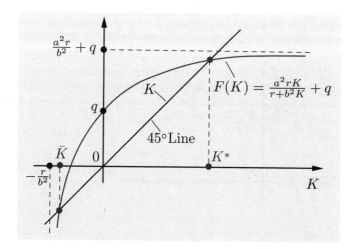

Figure 4.1.1 Graphical construction of the solutions of the Riccati equation (4.2)-(4.3) for the linear quadratic problem. The optimal cost function is $J^*(x) = K^*x^2$, where the scalar K^* solves the fixed point equation $K = F(K)$, with F being the function given by

$$F(K) = \frac{a^2 rK}{r + b^2 K} + q.$$

Because F is concave and monotonically increasing in the interval $(-r/b^2, \infty)$ and "flattens out" as $K \to \infty$, as shown in the figure, the quadratic Riccati equation $K = F(K)$ has one positive solution K^* and one negative solution, denoted $\bar{K}$.

nonnegative real line. This equation has another solution, denoted by $\bar{K}$ in Fig. 4.1.1, which lies within the negative real line and is of no interest. The optimal policy is a linear function of the state and has the form

$$\mu^*(x) = L^*x,$$

where L^* is the scalar given by

$$L^* = -\frac{abK^*}{r + b^2 K^*}. \tag{4.4}$$

4.2 COST FUNCTIONS OF STABLE LINEAR POLICIES

Suppose that we are given a linear policy of the form

$$\mu(x) = Lx,$$

where L is a scalar. The corresponding closed loop system is

$$x_{k+1} = (a + bL)x_k = (a + bL)^k x_0,$$

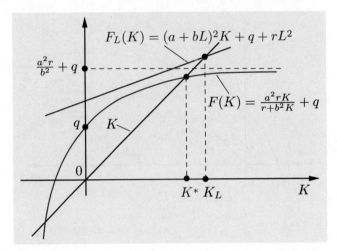

Figure 4.1.2 Illustration of the construction of the cost function of a linear policy $\mu(x) = Lx$, which is stable, i.e., $|a + bL| < 1$. The cost function $J_\mu(x)$ has the form $J_\mu(x) = K_L x^2$, where K_L is the unique solution of the linear equation $K = F_L(K)$, where
$$F_L(K) = (a + bL)^2 K + q + rL^2,$$
is the Riccati equation operator corresponding to L (i.e., the analog of T_μ). If μ is unstable, we have $J_\mu(x) = \infty$ for all $x \neq 0$.

and the cost $J_\mu(x_0)$ is calculated as

$$\sum_{k=0}^{\infty} \left(q(a+bL)^{2k} x_0^2 + rL^2 (a+bL)^{2k} x_0^2 \right) = \lim_{N \to \infty} \sum_{k=0}^{N-1} (q + rL^2)(a+bL)^{2k} x_0^2.$$

Assuming $|a + bL| < 1$, i.e., that the closed loop system is stable, the above summation yields
$$J_\mu(x) = K_L x^2,$$
for every initial state x, where

$$K_L = \frac{q + rL^2}{1 - (a + bL)^2}. \tag{4.5}$$

If on the other hand, $|a + bL| \geq 1$, i.e., the closed loop system is unstable, the summation yields $J_\mu(x_0) = \infty$ for all $x_0 \neq 0$.

It can be seen with a straightforward calculation that K_L is the unique solution of the linear equation

$$K = F_L(K), \tag{4.6}$$

where

$$F_L(K) = (a + bL)^2 K + q + rL^2; \tag{4.7}$$

Sec. 4.2   Cost Functions of Stable Linear Policies

see Fig. 4.1.2. Again, by replacing the Bellman operator T_μ of the stable policy $\mu(x) = Lx$ with the Riccati equation operator F_L, we can analyze the action of T_μ on the subspace of quadratic functions $J(x) = Kx^2$. Note that when $|a + bL| > 1$, so that μ is unstable, we have $J_\mu(x) = \infty$ for all $x \neq 0$, and the graph of F_L intersects the 45-degree line at a negative K. Then the equation $K = F_L(K)$ has the negative solution

$$\frac{q + rL^2}{1 - (a + bL)^2},$$

but this solution is unrelated to the cost function $J_\mu(\cdot)$, which has infinite value for all $x \neq 0$.

We summarize the Riccati equation formulas and the relation between linear policies of the form $\mu(x) = Lx$ and their quadratic cost functions in the following table.

Riccati Equation Formulas for One-Dimensional Problems

Riccati equation for minimization [cf. Eqs. (4.2) and (4.3)]

$$K = F(K), \qquad F(K) = \frac{a^2 r K}{r + b^2 K} + q.$$

Riccati equation for a stable linear policy $\mu(x) = Lx$ [cf. Eqs. (4.6) and (4.7)]

$$K = F_L(K), \qquad F_L(K) = (a + bL)^2 K + q + rL^2.$$

Gain L_K of lookahead linear policy associated with K [cf. Eq. (4.4)]

$$L_K = -\frac{abK}{r + b^2 K}.$$

Cost coefficient K_L of linear policy $\mu(x) = Lx$ [cf. Eq. (4.5)]

$$K_L = \frac{q + rL^2}{1 - (a + bL)^2}.$$

The one-dimensional problem of this chapter is well suited for geometric interpretations such as the ones we gave earlier in the preceding chapter, because approximation in value space, and the VI, rollout, and PI algorithms, involve quadratic cost functions $J(x) = Kx^2$, which can be represented by one-dimensional graphs as functions of just the number K. In particular, Bellman's equation can be replaced by the Riccati equation (4.3). Similarly, the figures in Chapter 3 for approximation in value space with one-step and multistep lookahead, the region of stability, rollout, and PI can be represented by one-dimensional graphs. We will next present these graphs and obtain corresponding geometrical insights. Note that our discussion applies qualitatively to multidimensional linear quadratic problems, and can be verified to a great extent by analysis, but an effective geometrical illustration is only possible when the system is one-dimensional.

4.3 VALUE ITERATION

The VI algorithm for the one-dimensional linear quadratic problem is illustrated in Fig. 4.3.1. It has the form

$$K_{k+1} = F(K_k);$$

cf. Example 2.1.1. As can be seen from the figure, the algorithm converges to K^* starting from anywhere in the interval $(\bar{K}, \infty)$, where $\bar{K}$ is the negative solution. In particular, the algorithm converges to K^* starting from any nonnegative value of K.

It is interesting to note that, starting from values K_0 with $K_0 \leq \bar{K}$, the algorithm does not converge to the optimal K^*. From Fig. 4.3.1, it can be seen that if

$$-\frac{r}{b^2} < K_0 \leq \bar{K},$$

it converges to the negative solution $\bar{K}$. The threshold $-r/b^2$ is the asymptotic value to the left of which $F(K)$ drops to $-\infty$. When

$$K_0 \leq -\frac{r}{b^2},$$

for the corresponding function $J_0(x) = K_0 x^2$ we have $(TJ_0)(x) = -\infty$ for all x, and the algorithm is not well defined. The literature on linear quadratic problems universally assumes that the iterative solution of the Riccati equation is started with nonnegative K_0, since to select a negative starting K_0 makes little sense. Note, however, that the nice behavior of VI just described depends on the positivity of the state cost coefficient q. In Section 4.8, we will discuss the exceptional case where $q = 0$.

4.4 ONE-STEP AND MULTISTEP LOOKAHEAD - NEWTON STEP INTERPRETATIONS

In this section, we consider approximation in value space with a quadratic terminal cost function $\tilde{J}(x) = \tilde{K}x^2$; cf. Fig. 4.4.1. We will express the cost function $J_{\tilde{\mu}}$ of the corresponding one-step lookahead policy $\tilde{\mu}$ in terms of

Sec. 4.4   One-Step and Multistep Lookahead - Newton Step   87

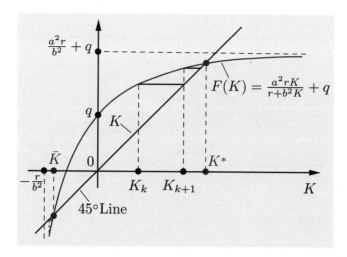

Figure 4.3.1 Graphical illustration of value iteration for the linear quadratic problem. It has the form $K_{k+1} = F(K_k)$ where

$$F(K) = \frac{a^2 r K}{r + b^2 K} + q.$$

It is essentially equivalent to the VI algorithm with a quadratic starting function

$$J_0(x) = K_0 x^2.$$

The algorithm converges to K^* starting from anywhere in the interval $(\bar{K}, \infty)$, where $\bar{K}$ is the negative solution, as shown in the figure. Starting from values K_0 with

$$-\frac{r}{b^2} < K_0 \leq \bar{K},$$

the algorithm converges to the negative solution $\bar{K}$. When

$$K_0 \leq -\frac{r}{b^2},$$

we have $(TJ_0)(x) = -\infty$ for all x, and the algorithm is undefined.

$\tilde{K}$ (assuming that $\tilde{K}$ belongs to the region of stability), and we will prove that the transformation from $\tilde{J}$ to $J_{\tilde{\mu}}$ is equivalent to a Newton step for solving the Riccati equation starting from $\tilde{K}$.

In particular, for a linear quadratic problem, the one-step lookahead policy is given by

$$\tilde{\mu}(x) \in \arg\min_u \left[qx^2 + ru^2 + \tilde{K}(ax + bu)^2 \right],$$

which after a straightforward calculation, yields

$$\tilde{\mu}(x) = \tilde{L}x,$$

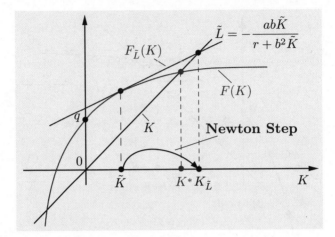

Figure 4.4.1 Illustration of approximation in value space with one-step lookahead for the linear quadratic problem. Given a terminal cost approximation $\tilde{J} = \tilde{K}x^2$, we compute the corresponding linear policy $\tilde{\mu}(x) = \tilde{L}x$, where

$$\tilde{L} = -\frac{ab\tilde{K}}{r + b^2\tilde{K}},$$

and the corresponding cost function $K_{\tilde{L}}x^2$, using the Newton step shown.

with the linear policy coefficient given by

$$\tilde{L} = -\frac{ab\tilde{K}}{r + b^2\tilde{K}}.$$

Note, however, that this policy will not be stable if $|a + b\tilde{L}| \geq 1$, or equivalently if

$$\left| a - \frac{ab^2\tilde{K}}{r + b^2\tilde{K}} \right| \geq 1.$$

We may also construct the linearization of the function F at $\tilde{K}$, and solve the corresponding linearized problem with a Newton step, as illustrated in Fig. 4.4.1. The case of ℓ-step lookahead minimization can be similarly interpreted. Instead of linearizing F at $\tilde{K}$, we linearize at $K_{\ell-1} = F^{\ell-1}(\tilde{K})$, i.e., at the result of $\ell - 1$ successive applications of F starting with $\tilde{K}$. Figure 4.4.2 depicts the case $\ell = 2$.

An important consequence of the Newton step interpretation is a classical quadratic convergence rate result: there exists an open interval containing K^* and a constant $c > 0$ such that for all $\tilde{K}$ within the open interval we have

$$|K_{\tilde{L}} - K^*| \leq c|\tilde{K} - K^*|^2, \tag{4.8}$$

Sec. 4.4 One-Step and Multistep Lookahead - Newton Step

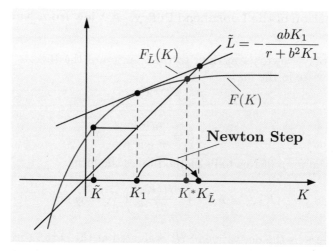

Figure 4.4.2 Illustration of approximation in value space with two-step lookahead for the linear quadratic problem. Starting with a terminal cost approximation $\tilde{J} = \tilde{K}x^2$, we obtain K_1 using a single value iteration. We then compute the corresponding linear policy $\tilde{\mu}(x) = \tilde{L}x$, where

$$\tilde{L} = -\frac{abK_1}{r + b^2 K_1}$$

and the corresponding cost function $K_{\tilde{L}}x^2$, using the Newton step shown.

where $\tilde{L}$ corresponds to the policy obtained by approximation in value space with one-step lookahead and terminal cost approximation $\tilde{J}(x) = \tilde{K}x^2$, so that

$$\tilde{L} = -\frac{ab\tilde{K}}{r + b^2 \tilde{K}}, \qquad (4.9)$$

and

$$K_{\tilde{L}} = \frac{q + r\tilde{L}^2}{1 - (a + b\tilde{L})^2}. \qquad (4.10)$$

Figure 4.4.2 also suggests another result that stems from the concavity property of the Riccati equation, namely that if $\hat{K}$ is a scalar that lies strictly within the region of stability, i.e.,

$$\left| a - \frac{ab^2 \hat{K}}{r + b^2 \hat{K}} \right| < 1,$$

then there exists a constant $c > 0$ such that all $\tilde{K} \geq \hat{K}$ satisfy Eq. (4.8). This result will not be proved here, but follows from the line of convergence analysis of Newton's method, which is given in the Appendix.

We will next show the quadratic convergence rate estimate (4.8) by verifying that $K_{\tilde{L}}$ *is the result of a Newton step for solving the Riccati equation* $K = F(K)$ *starting from* $\tilde{K}$.

Cost Function of the Lookahead Policy - A View from Newton's Method

We will apply Newton's method to the solution of the Riccati Eq. (4.3), which we write in the form
$$H(K) = 0,$$
where
$$H(K) = K - \frac{a^2 r K}{r + b^2 K} - q. \tag{4.11}$$

The classical form of Newton's method takes the form
$$K_{k+1} = K_k - \left(\frac{\partial H(K_k)}{\partial K}\right)^{-1} H(K_k), \tag{4.12}$$

where $\frac{\partial H(K_k)}{\partial K}$ is the derivative of H, evaluated at the current iterate K_k.

We will show analytically that the operation that generates K_L starting from K is a Newton iteration of the form (4.12) (an alternative is to argue graphically, as in Fig. 3.2.1). In particular, we will show (simplifying notation by skipping tilde) that for all K that lead to a stable one-step lookahead policy, we have

$$K_L = K - \left(\frac{\partial H(K)}{\partial K}\right)^{-1} H(K), \tag{4.13}$$

where we denote by
$$K_L = \frac{q + rL^2}{1 - (a + bL)^2} \tag{4.14}$$

the quadratic cost coefficient of the one-step lookahead linear policy $\mu(x) = Lx$ corresponding to the cost function approximation $J(x) = Kx^2$:

$$L = -\frac{abK}{r + b^2 K} \tag{4.15}$$

[cf. Eqs. (4.9) and (4.10)].

Our approach for showing the Newton step formula (4.13) is to express each term in this formula in terms of L, and then show that the formula holds as an identity for all L. To this end, we first note from Eq. (4.15) that K can be expressed in terms of L as

$$K = -\frac{rL}{b(a + bL)}. \tag{4.16}$$

Furthermore, by using Eqs. (4.15) and (4.16), $H(K)$ as given in Eq. (4.11) can be expressed in terms of L as follows:

$$H(K) = -\frac{rL}{b(a + bL)} + \frac{arL}{b} - q. \tag{4.17}$$

Sec. 4.5   Sensitivity Issues   91

Moreover, by differentiating the function H of Eq. (4.11), we obtain after a straightforward calculation

$$\frac{\partial H(K)}{\partial K} = 1 - \frac{a^2 r^2}{(r+b^2 K)^2} = 1 - (a+bL)^2, \qquad (4.18)$$

where the second equation follows from Eq. (4.15). Having expressed all the terms in the Newton step formula (4.13) in terms of L through Eqs. (4.14), (4.16), (4.17), and (4.18), we can write this formula in terms of L only as

$$\frac{q + rL^2}{1 - (a+bL)^2} = -\frac{rL}{b(a+bL)} - \frac{1}{1-(a+bL)^2}\left(-\frac{rL}{b(a+bL)} + \frac{arL}{b} - q\right),$$

or equivalently as

$$q + rL^2 = -\frac{rL\bigl(1 - (a+bL)^2\bigr)}{b(a+bL)} + \frac{rL}{b(a+bL)} - \frac{arL}{b} + q.$$

A straightforward calculation now shows that this equation holds as an identity for all L.

We have thus shown that K_L is related to K by the Newton step formula (4.13) for all K. Consequently, from classical calculus results on Newton's method, the quadratic convergence rate estimate (4.8) follows.

In the case of ℓ-step lookahead this result takes a stronger form, whereby the analog of the quadratic convergence rate estimate (4.8) has the form

$$|K_{\tilde{L}} - K^*| \le c\bigl|F^{\ell-1}(\tilde{K}) - K^*\bigr|^2,$$

where $F^{\ell-1}(\tilde{K})$ is the result of the $(\ell-1)$-fold application of the mapping F to $\tilde{K}$. Thus a stronger bound for $|K_{\tilde{L}} - K^*|$ is obtained.

4.5 SENSITIVITY ISSUES

An interesting consequence of the Newton step relation (4.13) between K and K_L relates to sensitivity to changes in K. In particular, we will show that asymptotically, near K^*, small changes in K induce much smaller changes in K_L. Mathematically, given K_1 and K_2 that lie within the region of stability, so they lead to stable corresponding one-step lookahead policies $\mu_1(x) = L_1 x$ and $\mu_2(x) = L_2 x$, we have

$$\frac{|K_{L_1} - K_{L_2}|}{|K_1 - K_2|} \to 0 \quad \text{as } |K_1 - K_2| \to 0 \text{ and } H(K_2) \to 0, \qquad (4.19)$$

as we will show next.

This result also holds for multidimensional linear quadratic problems, and also holds in various forms, under appropriate conditions, for more general problems. The practical significance of this result (and its extension to the case of a general differentiable Bellman operator T) is that near K^* (or J^*, in the general case), *small off-line training changes (i.e., small changes in K, or $\tilde{J}$, respectively) are rendered by the Newton step relatively insignificant as far as their effect on-line play performance is concerned (i.e., changes in K_L, or $J_{\tilde{\mu}}$, respectively)*. Small off-line training changes may result from applying alternative cost function approximation methods, which rely on similarly powerful feature-based or neural network-based architectures (e.g., different forms of temporal difference methods, aggregation methods, approximate linear programming, etc.).

To see why the sensitivity Eq. (4.19) holds, we rewrite the Newton iteration formula

$$K_{L_1} = K_1 - \left(\frac{\partial H(K_1)}{\partial K}\right)^{-1} H(K_1),$$

[cf. Eq. (4.13)] by using the first order Taylor approximations†

$$\left(\frac{\partial H(K_1)}{\partial K}\right)^{-1} = \left(\frac{\partial H(K_2)}{\partial K}\right)^{-1} + O(|K_1 - K_2|),$$

$$H(K_1) = H(K_2) + \frac{\partial H(K_2)}{\partial K}(K_1 - K_2) + o(|K_1 - K_2|).$$

We obtain

$$K_{L_1} = K_1 - \left(\left(\frac{\partial H(K_2)}{\partial K}\right)^{-1} + O(|K_1 - K_2|)\right)$$
$$\left(H(K_2) + \frac{\partial H(K_2)}{\partial K}(K_1 - K_2) + o(|K_1 - K_2|)\right),$$

which yields

$$K_{L_1} = K_1 - \left(\frac{\partial H(K_2)}{\partial K}\right)^{-1} H(K_2) - (K_1 - K_2) + O(|K_1 - K_2|)H(K_2)$$
$$+ o(|K_1 - K_2|).$$

---

† We are using here the standard calculus notation whereby $O(|K_1 - K_2|)$ denotes a function of (K_1, K_2) such that $O(|K_1 - K_2|) \to 0$ as $|K_1 - K_2| \to 0$, and $o(|K_1 - K_2|)$ denotes a function of (K_1, K_2) such that $o(|K_1 - K_2|)/|K_1 - K_2| \to 0$ as $|K_1 - K_2| \to 0$.

Sec. 4.5 Sensitivity Issues

The preceding equation, together with the Newton iteration formula [cf. Eq. (4.13)]

$$K_{L_2} = K_2 - \left(\frac{\partial H(K_2)}{\partial K}\right)^{-1} H(K_2),$$

yields

$$K_{L_1} = K_{L_2} + O(|K_1 - K_2|)H(K_2) + o(|K_1 - K_2|), \qquad (4.20)$$

or

$$\frac{K_{L_1} - K_{L_2}}{|K_1 - K_2|} = \frac{O(|K_1 - K_2|)}{|K_1 - K_2|} H(K_2) + \frac{o(|K_1 - K_2|)}{|K_1 - K_2|}. \qquad (4.21)$$

The first term of the right hand side above tends to 0 as $H(K_2) \to 0$, while the second term tends to 0 as $|K_1 - K_2| \to 0$. This proves the desired sensitivity Eq. (4.19). Also by tracing through the preceding calculations, it can be seen that the term $O(|K_1 - K_2|)$ multiplying $H(K_2)$ in Eq. (4.20) is equal to

$$\left(\frac{\partial H(K_1)}{\partial K}\right)^{-1} - \left(\frac{\partial H(K_2)}{\partial K}\right)^{-1}, \qquad (4.22)$$

and is close to 0 if $H(K)$ is nearly linear. Figure 4.5.1 illustrates the sensitivity estimate (4.21).

Note that from Eq. (4.21), the ratio $(K_{L_1} - K_{L_2})/|K_1 - K_2|$ depends on how close K_2 is to K^*, i.e., on the size of $H(K_2)$. In particular, if $K_2 = K^*$, we recover the superlinear convergence rate

$$\frac{|K_{L_1} - K^*|}{|K_1 - K^*|} = \frac{o(|K_1 - K^*|)}{|K_1 - K^*|}.$$

On the other hand, if $H(K_2)$ is far from 0 and H has large second derivatives near K_1 and K_2 [so that the difference of inverse derivatives (4.22) is large], the ratio $(K_{L_1} - K_{L_2})/|K_1 - K_2|$ can be quite large even if $|K_1 - K_2|$ is rather small. This sometimes tends to happen when K_1 and K_2 are close to the boundary of the region of stability, in which case it is important to bring the effective start of the Newton step closer to K^*, possibly by using multistep lookahead and/or truncated rollout with a stable policy.

The preceding derivation can be extended to hold for a general case of a mapping T as long as T is differentiable. If T is nondifferentiable at J^* the derivation breaks down and a sensitivity result such as the one of Eq. (4.19) may not hold in exceptional cases. Note, however, that in the case of a discounted problem with finite numbers of states and controls, where $(TJ)(x)$ is a piecewise linear function of J for every x, a stronger result can be proved: there is a sphere centered at J^* such that if $\tilde{J}$ lies within this sphere, the one-step lookahead policy corresponding to $\tilde{J}$ is optimal (see the book [Ber20a], Prop. 5.5.2).

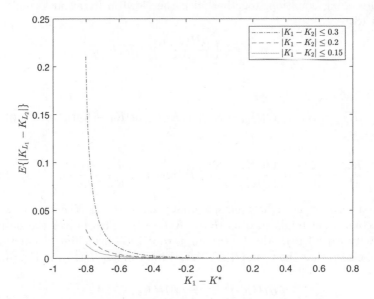

Figure 4.5.1 Illustration of the sensitivity estimate (4.21) for the case $a = 1$, $b = 2$, $q = 1$, $r = 0.5$. The difference $K_1 - K_2$ is denoted by ϵ. The figure shows the expected value of $|K_{L_1} - K_{L_2}|$ as the distance $K_1 - K^*$ varies, and as K_2 is selected randomly according to a uniform probability distribution in the interval $[K_1 - \epsilon, K_1 + \epsilon]$, for three different values of ϵ (0.15, 0.2, and 0.3). Note that for $K_1 > K^*$, the change $|K_{L_1} - K_{L_2}|$ is much smaller than for $K_1 < K^*$, because the difference of inverse derivatives (4.22) is small for $K_1 > K^*$, even for large values of ϵ (H is nearly linear).

4.6 ROLLOUT AND POLICY ITERATION

The rollout algorithm with a stable base policy μ is illustrated in Fig. 4.6.1. The PI algorithm is simply the repeated application of rollout. Let us derive the algorithm starting from a linear base policy of the form

$$\mu^0(x) = L_0 x,$$

where L_0 is a scalar. We require that L_0 is such that the closed loop system

$$x_{k+1} = (a + bL_0)x_k, \tag{4.23}$$

is stable, i.e., $|a + bL_0| < 1$. This is necessary for the policy μ^0 to keep the state bounded and the corresponding costs $J_{\mu^0}(x)$ finite. We will see that the PI algorithm generates a sequence of linear stable policies.

To describe the policy evaluation and policy improvement phases for the starting policy μ^0, we first calculate J_{μ^0} by noting that it involves the uncontrolled closed loop system (4.23) and a quadratic cost function. Similar to our earlier calculations, it has the form

$$J_{\mu^0}(x) = K_0 x^2, \tag{4.24}$$

Sec. 4.6 Rollout and Policy Iteration

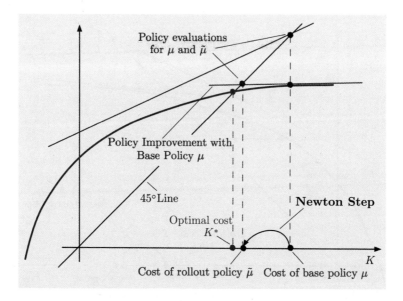

Figure 4.6.1 Illustration of rollout and policy iteration for the linear quadratic problem.

where
$$K_0 = \frac{q + rL_0^2}{1 - (a + bL_0)^2}. \tag{4.25}$$

Thus, the policy evaluation phase of PI for the starting linear policy $\mu^0(x) = L_0 x$ yields J_{μ^0} in the form (4.24)-(4.25). The policy improvement phase involves the quadratic minimization
$$\mu^1(x) \in \arg\min_u \left[qx^2 + ru^2 + K_0(ax + bu)^2 \right],$$
and after a straightforward calculation yields μ^1 as the linear policy $\mu^1(x) = L_1 x$, where
$$L_1 = -\frac{abK_0}{r + b^2 K_0}.$$
It can also be verified that μ^1 is a stable policy. An intuitive way to get a sense of this is via the cost improvement property of PI: we have $J_{\mu^1}(x) \leq J_{\mu^0}(x)$ for all x, so $J_{\mu^1}(x)$ must be finite, which implies stability of μ^1.

The preceding calculation can be continued, so the PI algorithm yields the sequence of stable linear policies
$$\mu^k(x) = L_k x, \qquad k = 0, 1, \ldots,$$
where L_{k+1} is generated by the iteration
$$L_{k+1} = -\frac{abK_k}{r + b^2 K_k},$$

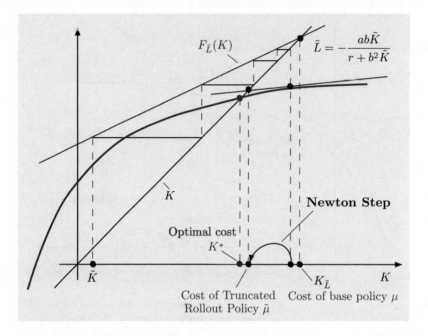

Figure 4.6.2 Illustration of truncated rollout with a stable base policy $\mu(x) = Lx$ and terminal cost approximation $\tilde{K}$ for the linear quadratic problem. In this figure the number of rollout steps is $m = 4$, and we use one-step lookahead minimization.

with K_k given by
$$K_k = \frac{q + rL_k^2}{1 - (a + bL_k)^2},$$
[cf. Eq. (4.25)].

The corresponding cost function sequence has the form $J_{\mu^k}(x) = K_k x^2$. Part of the classical linear quadratic theory is that J_{μ^k} converges to the optimal cost function J^*, while the generated sequence of linear policies $\{\mu^k\}$, where $\mu^k(x) = L_k x$, converges to the optimal policy. The convergence rate of the sequence $\{K_k\}$ is quadratic, as shown earlier, i.e., there exists a constant c such that
$$|K_{k+1} - K^*| \le c|K_k - K^*|^2,$$
for all k, assuming that the initial policy is linear and stable. This result was proved by Kleinman [Kle68] for the continuous time version of the linear quadratic problem, and it was extended later to more general problems; see the references given in Section 3.3 and the book [Ber20a], Chapter 1. For a recent proof of the quadratic convergence for linear discrete-time quadratic problems, see Lopez, Alsalti, and Muller [LAM21].

4.7 TRUNCATED ROLLOUT - LENGTH OF LOOKAHEAD ISSUES

Truncated rollout with a stable linear base policy $\mu(x) = Lx$ and terminal cost approximation $\tilde{J}(x) = \tilde{K}x^2$ is illustrated in Fig. 4.6.2. The rollout policy $\tilde{\mu}$ is obtained from the equation

$$T_{\tilde{\mu}} T^{\ell-1} T_\mu^m \tilde{J} = T^\ell T_\mu^m \tilde{J},$$

where $\ell \geq 1$ is the length of the lookahead minimization and $m \geq 0$ is the length of the rollout lookahead, with $m = 0$ corresponding to no lookahead by rollout (in Fig. 4.6.2, we have $\ell = 1$ and $m = 4$).

We mentioned some interesting performance issues in our discussion of truncated rollout in Section 3.4, and we will now revisit these issues within the context of our linear quadratic problem. In particular we noted that:

(a) Lookahead by rollout with a stable policy has a beneficial effect on the stability properties of the lookahead policy.

(b) Lookahead by rollout may be an economic substitute for lookahead by minimization, in the sense that it may achieve a similar performance for the truncated rollout policy at significantly reduced computational cost.

These statements are difficult to establish analytically in some generality. However, they can be intuitively understood in the context with our one-dimensional linear quadratic problem, using geometrical constructions like the one of Fig. 4.6.2.

In particular, let us consider our one-dimensional linear quadratic problem, the corresponding value K^* (the optimal cost coefficient), and the value K_s, which demarcates the region of stability, i.e., one-step lookahead yields a stable policy if and only if $\tilde{K} > K_s$. Consider also two parameters of the truncated rollout method: K_μ (the cost coefficient of the base policy), and $\tilde{K}$ (the terminal cost approximation coefficient).

We have $K_s \leq K^* \leq K_\mu$, so the three parameters K_s, K^*, K_μ divide the real line in the four intervals I through IV, depicted in Fig. 4.7.1. Then by examining Fig. 4.6.2, we see that the behavior of truncated rollout depends on the interval in which the terminal cost coefficient $\tilde{K}$ lies. In particular:

(a) *For $\tilde{K}$ in interval I*: Long total $(\ell+m)$-step lookahead is needed to get the starting point of the Newton step within the region of stability. Best and computationally economical results are obtained by taking $\ell = 1$ and m large enough to bring the starting point of the Newton step within the region of stability, and hopefully close to K^*.

(b) *For $\tilde{K}$ in interval II*: $\ell = 1$ and $m \geq 0$ are sufficient for stability. Best results are obtained when $\ell = 1$ and m is the (generally unknown) value that brings the starting point of the Newton step close to K^*.

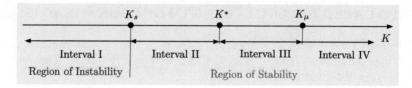

Figure 4.7.1 Illustration of the behavior of truncated rollout for the linear quadratic problem. We consider the four intervals I, II, III, and IV, defined by the boundary of the region of stability K_s, the optimal cost coefficient K^*, and the cost coefficient K_μ of the base policy μ. Using rollout with base policy μ, number of rollout steps $m > 0$, and a terminal cost coefficient $\tilde{K}$ in intervals I, II, and IV improves the stability guarantee and/or the performance of the lookahead policy $\tilde{\mu}$ over the case of no rollout, i.e., $m = 0$. In the case where $\tilde{K}$ lies in interval III, using $m > 0$ rather $m = 0$ deteriorates somewhat the performance of the lookahead policy $\tilde{\mu}$, but still maintains the cost improvement property $K_{\tilde{\mu}} \leq K_\mu$.

(c) *For $\tilde{K}$ in interval III*: $\ell = 1$ and $m \geq 0$ are sufficient for stability. Best results are obtained when $\ell = 1$ and $m = 0$ (since the rollout lookahead is counterproductive and takes the starting point of the Newton step away from K^* and towards K_μ). Still, however, even with $m > 0$, we have the cost improvement property $K_{\tilde{\mu}} \leq K_\mu$.

(d) *For $\tilde{K}$ in interval IV*: $\ell = 1$ and $m \geq 0$ are sufficient for stability. Best results are obtained for values of m and ℓ, which depend on how far $\tilde{K}$ is from K_μ. Here, values that likely work well are the ones for which m is fairly large, and ℓ is close to 1 (large enough m will bring the starting point of the Newton step close to K_μ; $\ell = 1$ corresponds to the Newton step, $\ell > 1$ improves the starting point of the Newton step by value iteration, but is likely not worth the extra computational cost).

Of course, a practical difficulty here is that we don't know the interval in which $\tilde{K}$ lies. However, it is clear that by using rollout with $m \geq 1$ works well in most cases as an economical substitute for long lookahead minimization. In particular, when $\tilde{K}$ lies in intervals I, II, and IV, using $m > 0$ provides a stronger stability guarantee and improved performance through a better starting point for the Newton step. Even in the case where $\tilde{K}$ lies in interval III, using $m > 0$ is not very damaging: we still obtain performance that is no worse than the base policy, i.e., $K_{\tilde{\mu}} \leq K_\mu$. An interesting research question is to investigate analytically as well as computationally, the multidimensional analogs of the intervals I-IV (which will now become subsets of the set of symmetric matrices). While it seems that the preceding discussion should generalize broadly, an investigation of the multidimensional case promises to be both challenging and illuminating. It may also reveal exceptional behaviors, particularly when extensions to problems more general than linear quadratic are considered.

Sec. 4.8   *Exceptional Behavior in Linear Quadratic Problems*   99

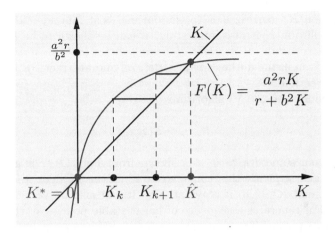

Figure 4.8.1 Illustration of the Bellman equation and the VI algorithm $K_{k+1} = F(K_k)$ for the linear quadratic problem in the exceptional case where $q = 0$.

4.8 EXCEPTIONAL BEHAVIOR IN LINEAR QUADRATIC PROBLEMS

It turns out that exceptional behavior can occur even for one-dimensional linear quadratic problems, when the positive definiteness assumption on the matrix Q is violated. In particular, let us consider the system

$$x_{k+1} = ax_k + bu_k, \tag{4.26}$$

and the cost function

$$\sum_{k=0}^{\infty} ru_k^2, \tag{4.27}$$

where a, b, r are scalars with $|a| > 1$, $b \neq 0$, $r > 0$. In this case there is no penalty for the state being nonzero (i.e., $q = 0$), while the system is unstable when left uncontrolled.

Here, since the cost per stage does not depend on the state, it is optimal to apply control $u = 0$ at any state x, i.e., $\mu^*(x) \equiv 0$, and the optimal cost function is $J^*(x) \equiv 0$. The Riccati equation is given by

$$K = F(K),$$

where F defined by

$$F(K) = \frac{a^2 rK}{r + b^2 K}.$$

As shown in Fig. 4.8.1, it has two nonnegative solutions:

$$K^* = 0 \quad \text{and} \quad \hat{K} = \frac{r(a^2 - 1)}{b^2}.$$

The solution K^* corresponds to the optimal cost function. It turns out that the solution $\hat{K}$ is also interesting: it can be shown to be *the optimal cost function within the subclass of linear policies that are stable*. A proof of this is given in the author's paper [Ber17c] and abstract DP monograph [Ber22a], Section 3.1.

Consider also the VI algorithm

$$K_{k+1} = F(K_k),$$

starting from some $K_0 > 0$. As shown from Fig. 4.8.1, it generates a positive scalar sequence that converges to $\hat{K}$. If the VI algorithm is started at the optimal $K^* = 0$, it stays at K^*. It can also be verified that the PI algorithm generates a sequence of linear stable policies starting from a linear stable policy. The sequence converges to the optimal stable policy that corresponds to $\hat{K}$.

In summary, the PI algorithm starting from a linear stable policy yields $\hat{J}$, the optimal cost function over the linear stable policies, but not the optimal cost function J^*. At the same time, the PI algorithm yields a a policy that is optimal over the linear stable policies, but not the optimal policy $\mu^*(x) \equiv 0$.

4.9 NOTES AND SOURCES

Linear quadratic problems are central in control theory, and have been the subject of extensive research. There are detailed accounts in most control theory textbooks, including Vol. I of the author's DP book [Ber17a].

The exceptional linear quadratic example of Section 4.8 provides an instance of a DP problem that exhibits a so-called "semicontractive behavior." By this we mean that some policies are "well-behaved" (stabilize the system in this case, or have Bellman operators that are contractive in other cases), while some other policies are not, and the VI algorithm tends to be attracted to the optimal cost function over the well-behaved policies only. Semicontractive models and their analysis are a major focal point of the abstract DP monograph [Ber22a], Chapters 3 and 4. They arise commonly in the context of stochastic shortest path problems, where some policies (called proper) are "well-behaved" in the sense that they guarantee that the termination state will be reached from every initial state, while other policies (called improper) are not.

5
Adaptive and Model Predictive Control

Contents

5.1. Systems with Unknown Parameters - Robust and PID
Control . . . . p. 102
5.2. Approximation in Value Space, Rollout, and Adaptive
Control . . . . p. 105
5.3. Approximation in Value Space, Rollout, and Model
Predictive Control . . . . p. 109
5.4. Terminal Cost Approximation - Stability Issues . . . . p. 112
5.5. Notes and Sources . . . . p. 118

In this chapter, we discuss some of the core control system design methodologies within the context of our approximation in value space framework. In particular, in the next two sections, we will discuss problems with unknown or changing problem parameters, and briefly review some of the principal types of adaptive control methods. We will then focus on schemes that are based on on-line replanning, including the use of rollout. The idea here is to use an approximation in value space scheme/Newton step in place of a full reoptimization of the controller, in response to the changed system parameters; we have noted this possibility in Chapter 1. Subsequently, in Sections 5.3 and 5.4, we will discuss the model predictive control methodology, and its connections with approximation in value space, Newton's method, adaptive control, and the attendant stability issues.

5.1 SYSTEMS WITH UNKNOWN PARAMETERS - ROBUST AND PID CONTROL

Our discussion so far dealt with problems with a known and unchanging mathematical model, i.e., one where the system equation, cost function, control constraints, and probability distributions of disturbances are perfectly known. The mathematical model may be available through explicit mathematical formulas and assumptions, or through a computer program that can emulate all of the mathematical operations involved in the model, including Monte Carlo simulation for the calculation of expected values. From our point of view, it makes no difference whether the mathematical model is available through closed form mathematical expressions or through a computer simulator: the methods that we discuss are valid either way, only their suitability for a given problem may be affected by the availability of mathematical formulas.

In practice, however, it is common that the system involves parameters that are either not known exactly or may change over time. In such cases it is important to design controllers that take the parameter changes into account. The methodology for doing so is generally known as *adaptive control*, an intricate and multifaceted subject, with many and diverse applications, and a long history.†

We should note also that unknown problem environments are an integral part of the artificial intelligence view of RL. In particular, to quote

---

† The difficulties of designing adaptive controllers are often underestimated. Among others, they complicate the balance between off-line training and on-line play, which we discussed in Chapter 1 in connection to AlphaZero. It is worth keeping in mind that as much as learning to play high quality chess is a great challenge, the rules of play are stable and do not change unpredictably in the middle of a game! Problems with changing system parameters can be far more challenging!

Sec. 5.1 *Systems with Unknown Parameters - Robust and PID Control* **103**

from the book by Sutton and Barto [SuB18], "learning from interaction with the environment is a foundational idea underlying nearly all theories of learning and intelligence." The idea of interaction with the environment is typically connected with the idea of exploring the environment to identify its characteristics. In control theory this is often viewed as part of the *system identification* methodology, which aims to construct mathematical models of dynamic systems by using data. The system identification process is often combined with the control process to deal with unknown or changing problem parameters. This is one of the most challenging areas of stochastic optimal and suboptimal control, and has been studied extensively since the early 1960s.

Robust and PID Control

Given a controller design that has been obtained assuming a nominal DP problem model, one possibility is to simply ignore changes in problem parameters. We may then try to design a controller that is adequate for the entire range of the changing parameters. This is sometimes called a *robust controller*. A robust controller makes no effort to keep track of changing problem parameters. It is just designed so that it is resilient to parameter changes, and in practice, it often tends to be biased towards addressing the worst case.

An important and time-honored robust control approach for continuous-state problems is the *PID (Proportional-Integral-Derivative) controller*; see e.g., the books by Aström and Hagglund [AsH95], [AsH06]. In particular, PID control aims to maintain the output of a single-input single-output dynamic system around a set point or to follow a given trajectory, as the system parameters change within a relatively broad range. In its simplest form, the PID controller is parametrized by three scalar parameters, which may be determined by a variety of methods, some of them manual/heuristic. PID control is used widely and with success, although its range of application is mainly restricted to relatively simple, single-input and single-output continuous-state control systems.

Combined System Identification and Control

In robust control schemes, such as PID control, no attempt is made to maintain a mathematical model and to track unknown model parameters as they change. Alternatively we may introduce into the controller a mechanism for measuring or estimating the unknown or changing system parameters, and make suitable control adaptations in response.†

---

† In the adaptive control literature, schemes that involve parameter estimation are sometimes called *indirect*, while schemes that do not involve parameter estimation (like PID control) are called *direct*. To quote from the book by Aström

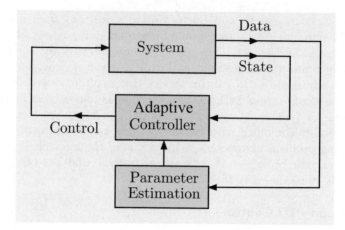

Figure 5.1.1 Schematic illustration of concurrent parameter estimation and system control. The system parameters are estimated on-line and the estimates are passed on to the controller whenever this is desirable (e.g., after the estimates change substantially). This structure is also known as indirect adaptive control.

Let us note here that updating problem parameters need not require an elaborate algorithm. In many cases the set of problem parameters may take a known finite set of values (for example each set of parameter values may correspond to a distinct maneuver of a vehicle, motion of a robotic arm, flying regime of an aircraft, etc). Once the control scheme detects a change in problem parameters, it can incorporate the change into the approximation in value space scheme, and in the case of policy rollout, it may switch to a corresponding predesigned base policy.

In what follows in this chapter (including our discussion of MPC in Section 5.3), we will assume that there is a mechanism to learn (perhaps imperfectly and by some unspecified procedure) the model of the system as it evolves over time. We will loosely refer to this learning process with the classical name *system identification*, but we will not go into specific identification methods, keeping in mind that such methods could be imprecise and challenging, but could also be fast and simple, depending on the problem at hand.

An apparently reasonable scheme is to separate the control process into two phases, a *system identification phase* and a *control phase*. In the first phase the unknown parameters are estimated, while the control takes no account of the interim results of estimation. The final parameter

---

and Wittenmark [AsW08], "indirect methods are those in which the estimated parameters are used to calculate required controller parameters" (see Fig. 5.1.1). The methods subsequently described in this section, and the rollout-based adaptive control methods discussed in the next section should be viewed as indirect.

estimates from the first phase are then used to implement an optimal or suboptimal policy in the second phase.

This alternation of estimation and control phases may be repeated several times during the system's operation in order to take into account subsequent changes of the parameters. Note that it is not necessary to introduce a hard separation between the identification and the control phases. They may be going on simultaneously, with new parameter estimates being generated in the background, and introduced into the control process, whenever this is thought to be desirable; see Fig. 5.1.1.

One drawback of this approach is that it is not always easy to determine when to terminate one phase and start the other. A second difficulty, of a more fundamental nature, is that the control process may make some of the unknown parameters invisible to the estimation process. This is known as the problem of *parameter identifiability*, which is discussed in the context of adaptive control in several sources. On-line parameter estimation algorithms, which address among others the issue of identifiability, have been discussed extensively in the control theory literature, but the corresponding methodology is complex and beyond our scope in this book. However, assuming that we can make the estimation phase work somehow, we are free to reoptimize the controller using the newly estimated parameters, in a form of on-line replanning process.

Unfortunately, there is still another difficulty with this type of on-line replanning: it may be hard to recompute an optimal or near-optimal policy on-line, using a newly identified system model. In particular, it may be impossible to use time-consuming and/or data-intensive methods that involve for example the training of a neural network, or discrete/integer control constraints. A simpler possibility is to use rollout, which we discuss in the next section.

5.2 APPROXIMATION IN VALUE SPACE, ROLLOUT, AND ADAPTIVE CONTROL

We will now consider an approach for dealing with unknown or changing parameters, which is based on rollout and on-line replanning. We have already noted this approach in Chapter 1, where we stressed the importance of fast on-line policy improvement.

Let us assume that some problem parameters change over time, while the controller estimates the changes on-line, perhaps after a suitable delay for data collection. The method by which the problem parameters are recalculated or become known is immaterial for the purposes of the following discussion. It may involve a limited form of parameter estimation, whereby the unknown parameters are "tracked" by data collection over a few time stages, with due attention paid to issues of parameter identifiability; or it may involve new features of the control environment, such as a changing number of servers and/or tasks in a service system.

We thus assume away/ignore the detailed issues of parameter estimation, and focus on revising the controller by on-line replanning based on the newly obtained parameters. This revision may be based on any suboptimal method, but rollout with some base policy is particularly attractive. The base policy may be either a fixed robust controller (such as some form of PID control) or it may be updated over time (in the background, on the basis of some unspecified rationale), in which case the rollout policy will be revised both in response to the changed base policy and in response to the changing parameters.

Here the advantage of rollout is that it is simple, reliable, and relatively fast. In particular, it does not require a complicated training procedure, based for example on the use of neural networks or other approximation architectures, so *no new policy is explicitly computed in response to the parameter changes*. Instead the available controls at the current state are compared through a one-step or multistep minimization, with cost function approximation provided by the base policy (cf. Fig. 5.2.1).

Another issue to consider is the stability and robustness properties of the rollout policy. In this connection, it can be generally proved, under mild conditions, that *if the base policy is stable within a range of parameter values, the same is true for the rollout policy*; this can also be inferred from Fig. 3.4.3. Related ideas have a long history in the control theory literature; see Beard [Bea95], Beard, Saridis, and Wen [BSW99], Jiang and Jiang [JiJ17], Kalise, Kundu, Kunisch [KKK20], Pang and Jiang [PaJ21].

The principal requirement for using rollout in an adaptive control context is that the rollout control computation should be fast enough to be performed between stages. In this connection, we note that accelerated/truncated or simplified versions of rollout, as well as parallel computation, can be used to meet this time constraint.

Generally, adaptive control by rollout and on-line replanning makes sense in situations where the calculation of the rollout controls for a given set of problem parameters is faster and/or more convenient than the calculation of the optimal controls for the same set of parameter values. These problems include cases involving nonlinear systems and/or difficult (e.g., integer) constraints.

The following example illustrates on-line replanning with the use of rollout in the context of the simple one-dimensional linear quadratic problem that we discussed earlier. The purpose of the example is to show analytically how rollout with a base policy that is optimal for a nominal set of problem parameters works well when the parameters change from their nominal values. This property is not practically useful in linear quadratic problems because when the parameter change, it is possible to calculate the new optimal policy in closed form, but it is indicative of the performance robustness of rollout in other contexts; for example linear quadratic problems with constraints.

Sec. 5.2 *Approximation in Value Space, Rollout, and Adaptive Control* **107**

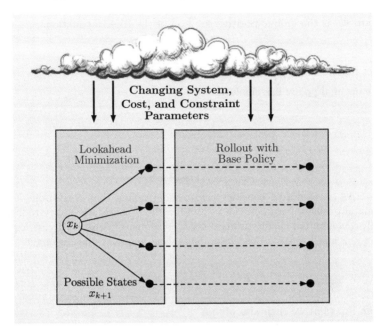

Figure 5.2.1 Schematic illustration of adaptive control by on-line replanning based on rollout. One-step lookahead minimization is followed by simulation with the base policy, which stays fixed. The system, cost, and constraint parameters are changing over time, and the most recent estimates of their values are incorporated into the lookahead minimization and rollout operations. Truncated rollout with multistep lookahead minimization and terminal cost approximation is also possible. The base policy may also be revised based on various criteria. For the discussion of this section, we may assume that all the changing parameter information is provided by some computation and sensor "cloud" that is beyond our control.

Example 5.2.1 (On-Line Replanning for Linear Quadratic Problems Based on Rollout)

Consider a deterministic undiscounted infinite horizon linear quadratic problem involving the linear system

$$x_{k+1} = x_k + bu_k,$$

and the quadratic cost function

$$\lim_{N\to\infty} \sum_{k=0}^{N-1} (x_k^2 + ru_k^2).$$

This is the one-dimensional problem of the preceding section for the special case where $a = 1$ and $q = 1$. The optimal cost function is given by

$$J^*(x) = K^* x^2,$$

where K^* is the unique positive solution of the Riccati equation

$$K = \frac{rK}{r+b^2K} + 1. \tag{5.1}$$

The optimal policy has the form

$$\mu^*(x) = L^*x, \tag{5.2}$$

where

$$L^* = -\frac{bK^*}{r+b^2K^*}. \tag{5.3}$$

As an example, consider the optimal policy that corresponds to the nominal problem parameters $b=2$ and $r=0.5$: this is the policy (5.2)-(5.3), with K computed as the positive solution of the quadratic Riccati Eq. (5.1) for $b=2$ and $r=0.5$. For these nominal parameter values, we have

$$K^* = \frac{2+\sqrt{6}}{4} \approx 1.11.$$

From Eq. (5.3) we then also obtain

$$L^* = -\frac{2+\sqrt{6}}{5+2\sqrt{6}}. \tag{5.4}$$

We will now consider changes of the values of b and r while keeping L constant to the preceding value, and we will compare the quadratic cost coefficients of the following three cost functions as b and r vary:

(a) The optimal cost function K^*x^2, where K^* is given by the positive solution of the Riccati Eq. (5.1).

(b) The cost function $K_L x^2$ that corresponds to the base policy

$$\mu_L(x) = Lx,$$

where L is given by Eq. (5.4). Here, we have (cf. Section 4.1)

$$K_L = \frac{1+rL^2}{1-(1+bL)^2}. \tag{5.5}$$

(c) The cost function $\tilde{K}_L x^2$ that corresponds to the rollout policy

$$\tilde{\mu}_L(x) = \tilde{L}x,$$

obtained by using the policy μ_L as base policy. Using the formulas derived earlier, we have [cf. Eq. (5.5)]

$$\tilde{L} = -\frac{bK_L}{r+b^2K_L},$$

and (cf. Section 4.1)

$$\tilde{K}_L = \frac{1 + r\tilde{L}^2}{1 - (1 + b\tilde{L})^2}.$$

Figure 5.2.2 shows the coefficients K^*, K_L, and $\tilde{K}_L$ for a range of values of r and b. We have

$$K^* \leq \tilde{K}_L \leq K_L.$$

The difference $K_L - K^*$ is indicative of the robustness of the policy μ_L, i.e., the performance loss incurred by ignoring the changes in the values of b and r, and continuing to use the policy μ_L, which is optimal for the nominal values $b = 2$ and $r = 0.5$, but suboptimal for other values of b and r. The difference $\tilde{K}_L - K^*$ is indicative of the performance loss due to using online replanning by rollout rather than using optimal replanning. Finally, the difference $K_L - \tilde{K}_L$ is indicative of the performance improvement due to online replanning using rollout rather than keeping the policy μ_L unchanged.

Note that Fig. 5.2.2 illustrates the behavior of the error ratio

$$\frac{\tilde{J} - J^*}{J - J^*},$$

where for a given initial state, $\tilde{J}$ is the rollout performance, J^* is the optimal performance, and J is the base policy performance. This ratio approaches 0 as $J - J^*$ becomes smaller because of the superlinear/quadratic convergence rate of Newton's method that underlies the rollout algorithm.

5.3 APPROXIMATION IN VALUE SPACE, ROLLOUT, AND MODEL PREDICTIVE CONTROL

In this section, we briefly discuss the MPC methodology, with a view towards its connection with approximation in value space and the rollout algorithm. We will focus on the undiscounted infinite horizon deterministic problem, which involves the system

$$x_{k+1} = f(x_k, u_k),$$

whose state x_k and control u_k are finite-dimensional vectors. The cost per stage is assumed nonnegative

$$g(x_k, u_k) \geq 0, \qquad \text{for all } (x_k, u_k),$$

(e.g., a positive definite quadratic cost). There are control constraints $u_k \in U(x_k)$, and to simplify the following discussion, we will initially consider no state constraints. We assume that the system can be kept at the origin at zero cost, i.e.,

$$f(0, \overline{u}_k) = 0, \quad g(0, \overline{u}_k) = 0 \qquad \text{for some control } \overline{u}_k \in U(0).$$

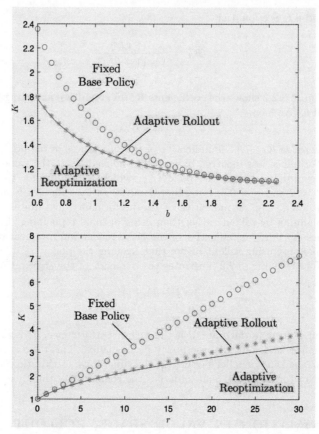

Figure 5.2.2 Illustration of control by rollout under changing problem parameters. The quadratic cost coefficients K^* (optimal, denoted by solid line), K_L (base policy, denoted by circles), and $\tilde{K}_L$ (rollout policy, denoted by asterisks) are shown for the two cases where $r = 0.5$ and b varies, and $b = 2$ and r varies. The value of L is fixed at the value that is optimal for $b = 2$ and $r = 0.5$ [cf. Eq. (5.4)]. The rollout policy performance is very close to optimal, even when the base policy is far from optimal.

Note that, as the figure illustrates, we have

$$\lim_{J \to J^*} \frac{\tilde{J} - J^*}{J - J^*} = 0,$$

where for a given initial state, $\tilde{J}$ is the rollout performance, J^* is the optimal performance, and J is the base policy performance. This is a consequence of the superlinear/quadratic convergence rate of Newton's method that underlies rollout, and guarantees that the rollout performance approaches the optimal much faster than the base policy performance does.

For a given initial state x_0, we want to obtain a sequence $\{u_0, u_1, \ldots\}$ that satisfies the control constraints, while minimizing the total cost.

Sec. 5.3 Approximation in Value Space - Model Predictive Control

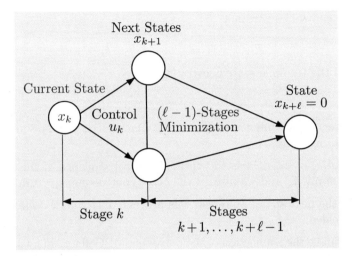

Figure 5.3.1 Illustration of the problem solved by a classical form of MPC at state x_k. We minimize the cost function over the next ℓ stages while imposing the requirement that $x_{k+\ell} = 0$. We then apply the first control of the optimizing sequence. In the context of rollout, the minimization over u_k is the one-step lookahead, while the minimization over $u_{k+1}, \ldots, u_{k+\ell-1}$ that drives $x_{k+\ell}$ to 0 is the base heuristic.

This is a classical problem in control system design, known as the *regulation problem*, where the aim is to keep the state of the system near the origin (or more generally some desired set point), in the face of disturbances and/or parameter changes. In an important variant of the problem, there are additional state constraints of the form $x_k \in X$, and there arises the issue of maintaining the state within X, not just at the present time but also in future times. We will address this issue later in this section.

The Classical Form of MPC - View as a Rollout Algorithm

We will first focus on a classical form of the MPC algorithm, proposed in the form given here by Keerthi and Gilbert [KeG88]. In this algorithm, at each encountered state x_k, we apply a control $\tilde{u}_k$ that is computed as follows; see Fig. 5.3.1:

(a) We solve an ℓ-stage optimal control problem involving the same cost function and the requirement that the state after ℓ steps is driven to 0, i.e., $x_{k+\ell} = 0$. This is the problem

$$\min_{u_t,\, t=k,\ldots,k+\ell-1} \sum_{t=k}^{k+\ell-1} g(x_t, u_t), \tag{5.6}$$

subject to the system equation constraints

$$x_{t+1} = f(x_t, u_t), \qquad t = k, \ldots, k+\ell-1, \tag{5.7}$$

the control constraints

$$u_t \in U(x_t), \qquad t = k, \ldots, k + \ell - 1, \tag{5.8}$$

and the terminal state constraint

$$x_{k+\ell} = 0. \tag{5.9}$$

Here ℓ is an integer with $\ell > 1$, which is chosen in some largely empirical way.

(b) If $\{\tilde{u}_k, \ldots, \tilde{u}_{k+\ell-1}\}$ is the optimal control sequence of this problem, we apply $\tilde{u}_k$ and we discard the other controls $\tilde{u}_{k+1}, \ldots, \tilde{u}_{k+\ell-1}$.

(c) At the next stage, we repeat this process, once the next state x_{k+1} is revealed.

To make the connection of the preceding MPC algorithm with rollout, we note that *the one-step lookahead function $\tilde{J}$ implicitly used by MPC [cf. Eq. (5.6)] is the cost function of a certain stable base policy*. This is the policy that drives to 0 the state after $\ell - 1$ stages (*not ℓ stages*) and keeps the state at 0 thereafter, while observing the state and control constraints, and minimizing the associated $(\ell-1)$-stages cost. This rollout view of MPC was first discussed in the author's paper [Ber05]. It is useful for making a connection with the approximate DP/RL, rollout, and its interpretation in terms of Newton's method. In particular, an important consequence is that *the MPC policy is stable*, since rollout with a stable base policy yields a stable policy, as we have discussed in Section 3.2.

We may also equivalently view the preceding MPC algorithm as rollout with $\bar{\ell}$-step lookahead, where $1 < \bar{\ell} < \ell$, with the base policy that drives to 0 the state after $\ell - \bar{\ell}$ stages and keeps the state at 0 thereafter. This suggests variations of MPC that involve truncated rollout with terminal cost function approximation, which we will discuss shortly.

Note also that when faced with changing problem parameters, it is natural to consider on-line replanning as per our earlier discussion. In particular, once new estimates of system and/or cost function parameters become available, MPC can adapt accordingly by introducing the new parameter estimates into the ℓ-stage optimization problem in (a) above.

5.4 TERMINAL COST APPROXIMATION - STABILITY ISSUES

In a common variant of MPC, the requirement of driving the system state to 0 in ℓ steps in the ℓ-stage MPC problem (5.6), is replaced by a nonnegative terminal cost $G(x_{k+\ell})$. Thus at state x_k, we solve the problem

$$\min_{u_t,\, t=k,\ldots,k+\ell-1} \left[G(x_{k+\ell}) + \sum_{t=k}^{k+\ell-1} g(x_t, u_t) \right], \tag{5.10}$$

Sec. 5.4   Terminal Cost Approximation - Stability Issues

instead of problem (5.6) where we require that $x_{k+\ell} = 0$. This variant can be viewed as rollout with one-step lookahead, and a base policy, which at state x_{k+1} applies the first control $\tilde{u}_{k+1}$ of the sequence $\{\tilde{u}_{k+1}, \ldots, \tilde{u}_{k+\ell-1}\}$ that minimizes

$$G(x_{k+\ell}) + \sum_{t=k+1}^{k+\ell-1} g(x_t, u_t).$$

It can also be viewed outside the context of rollout, as approximation in value space with ℓ-step lookahead minimization and terminal cost approximation given by G. Thus the preceding MPC controller may have cost function that is much closer to J^* than G is. This is due to the superlinear/quadratic convergence rate of Newton's method that underlies approximation in value space, as we have discussed in Chapter 3.

An important question is to choose the terminal cost approximation so that the resulting MPC controller is stable. Our discussion of Section 3.3 on the region of stability of approximation in value space schemes applies here. In particular, under the nonnegative cost assumption of this section, the MPC controller will be stable if $TG \leq G$ (using the abstract DP notation introduced in Chapter 3), or equivalently

$$(TG)(x) = \min_{u \in U(x)} \Big\{ g(x,u) + G\big(f(x,u)\big) \Big\} \leq G(x), \qquad \text{for all } x, \quad (5.11)$$

as noted in Section 3.2. This condition is sufficient for stability of the MPC controller, but it is not necessary. Figure 5.4.1 provides a graphical illustration. It shows that the condition $TG \leq G$ implies that $J^* \leq T^\ell G \leq T^{\ell-1}G$ for all $\ell \geq 1$ (the books [Ber12] and [Ber18a] provide mathematical proofs of this fact). This in turn implies that $T^\ell G$ lies within the region of the stability for all $\ell \geq 0$.

We also expect that as the length ℓ of the lookahead minimization is increased, the stability properties of the MPC controller are improved. In particular, *given $G \geq 0$, the resulting MPC controller is likely to be stable for ℓ sufficiently large*, since $T^\ell G$ ordinarily converges to J^*, which lies within the region of stability. Results of this type are known within the MPC framework under various conditions (see the papers by Mayne at al. [MRR00], Magni et al. [MDM01], the MPC book [RMD17], and the author's book [Ber20a], Section 3.1.2). Our discussion of stability in Sections 4.4 and 4.6 is also relevant within this context.

In another variant of MPC, in addition to the terminal cost function approximation G, we use truncated rollout, which involves running some stable base policy μ for a number of steps m; see Fig. 5.4.2. This is quite similar to standard truncated rollout, except that the computational solution of the lookahead minimization problem (5.10) may become complicated when the control space is infinite. As discussed in Section 3.3, *increasing the length of the truncated rollout enlarges the region of stability of the MPC controller*. The reason is that by increasing the length of the

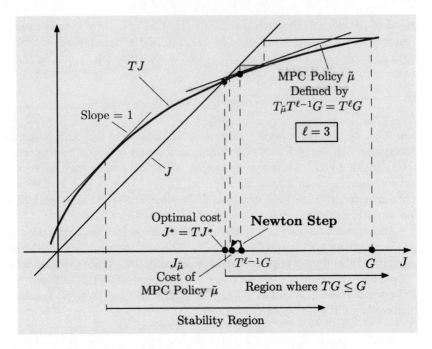

Figure 5.4.1 Illustration of the condition $TG \leq G$ or equivalently

$$(TG)(x) = \min_{u \in U(x)} \left\{ g(x,u) + G\big(f(x,u)\big) \right\} \leq G(x), \qquad \text{for all } x.$$

When satisfied by the terminal cost function approximation G, it guarantees the stability of the MPC policy $\tilde{\mu}$ with ℓ-step lookahead minimization, defined by

$$T_{\tilde{\mu}} T^{\ell-1} G = T^{\ell} G,$$

where for a generic policy μ, T_μ is defined (using the abstract DP notation of Chapter 3) by

$$(T_\mu J)(x) = g\big(x, \mu(x)\big) + J\big(f(x, \mu(x))\big), \qquad \text{for all } x.$$

In this figure, $\ell = 3$.

truncated rollout, we push the start of the Newton step towards of the cost function J_μ of the stable policy, which lies within the region of stability since $TJ_\mu \leq T_\mu J_\mu = J_\mu$; see also the discussion on linear quadratic problems in Section 4.7. The base policy may also be used to address state constraints; see the papers by Rosolia and Borelli [RoB17], [RoB19], and the discussion in the author's RL book [Ber20a].

Sec. 5.4 Terminal Cost Approximation - Stability Issues

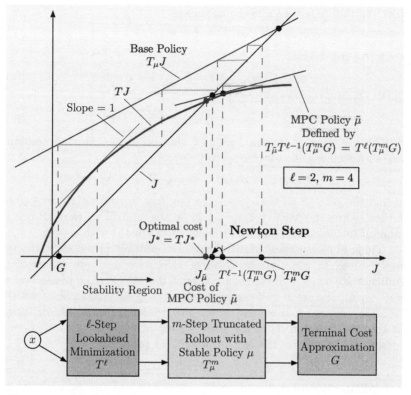

Figure 5.4.2 An MPC scheme with ℓ-step lookahead minimization, m-step truncated rollout with a stable base policy μ, and a terminal cost function approximation G, together with its interpretation as a Newton step. In this figure, $\ell = 2$ and $m = 4$. As m increases, $T_\mu^m G$ moves closer to J_μ, which lies within the region of stability.

A Rollout Variant of MPC with Multiple Terminal States and Base Policies

In another variation of MPC, proposed in the paper by Li et al. [LJM21], instead of driving the state to 0 at the end of ℓ steps, we consider multiple terminal system states at the end of the ℓ-step horizon, as well as the use of multiple base policies for rollout. In particular, in this scheme we have a finite set of states $\mathcal{X}$ and a finite set of stable base policies $\mathcal{M}$, and we assume that we have computed off-line the cost function values $J_\mu(x)$ for all $x \in \mathcal{X}$ and $\mu \in \mathcal{M}$. At state x_k, to compute the MPC control $\tilde{u}_k$, we solve for each $x \in \mathcal{X}$ a problem that is the same as the problem (5.6)-(5.9), which is solved by the classical form of MPC, except that the terminal state $x_{k+\ell}$ is equal to x instead of $x_{k+\ell} = 0$. This is the problem

$$\min_{u_t,\, t=k,\ldots,k+\ell-1} \sum_{t=k}^{k+\ell-1} g(x_t, u_t), \tag{5.12}$$

subject to the system equation constraints

$$x_{t+1} = f(x_t, u_t), \qquad t = k, \ldots, k+\ell-1, \tag{5.13}$$

the control constraints

$$u_t \in U(x_t), \qquad t = k, \ldots, k+\ell-1, \tag{5.14}$$

and the terminal state constraint

$$x_{k+\ell} = x. \tag{5.15}$$

Let $V(x_k; x)$ be the optimal value of this problem. Having computed $V(x_k; x)$ for all $x \in \mathcal{X}$, we compare all values

$$V(x_k; x) + J_\mu(x), \qquad x \in \mathcal{X}, \ \mu \in \mathcal{M},$$

and find the pair $(\overline{x}, \overline{\mu})$ that yields the minimal value of $V(x_k; x) + J_\mu(x)$. We then define the MPC control $\tilde{u}_k$ to be the control u_k that attains the minimum in the corresponding problem (5.12)-(5.15) with $x = \overline{x}$.

Thus, in this variant of MPC we solve multiple problems of the type that is solved in the classical form of MPC, for multiple values of the terminal state $x_{k+\ell}$, and we then compute the MPC control based on the "best" terminal state $x \in \mathcal{X}$, assuming that the "best" base policy $\overline{\mu}$ will be used after state $k+\ell$. It is possible to show, under appropriate conditions,† that the cost function $J_{\tilde{\mu}}$ of the MPC policy $\tilde{\mu}$, which applies $\tilde{\mu}(x_k) = \tilde{u}_k$ as described above, has the cost improvement property

$$J_{\tilde{\mu}}(x) \le J_\mu(x), \qquad \text{for all } x \in \mathcal{X}, \ \mu \in \mathcal{M}; \tag{5.16}$$

see [LJM21]. Moreover, based on this property and the assumption that the base policies $\mu \in \mathcal{M}$ are stable, it follows that the MPC policy $\tilde{\mu}$ thus obtained is also stable.

The preceding variation can also be used for systems with arbitrary state and control spaces, continuous as well as discrete. It is also well-suited for addressing state constraints, provided the base policies are designed to satisfy these constraints. In this case, the state constraints are included in the constraints of the ℓ-step problems (5.12)-(5.15). We refer to the paper [LJM21], which provides supporting analysis, extensions to the case where X is an infinite set, as well as computational results involving several types of problems, with both discrete and continuous state and control spaces.

We mention that the idea of using multiple base policies to evaluate the available controls at a given state, and selecting the control that yields the least cost, has been known since the original proposal of the paper [BTW97]. The main result for such schemes is a cost improvement property, whereby the rollout policy outperforms simultaneously all the base policies; cf. Eq. (5.16). This property is also discussed in Sections 6.3 and 6.4, as well as the books [Ber17a], [Ber19a], [Ber20a].

---

† These conditions include that for every $x \in \mathcal{X}$, we have $f(x, \mu(x)) \in \mathcal{X}$ for some $\mu \in \mathcal{X}$, which plays the same role as the assumption that the origin is cost free and absorbing in the classical form of MPC.

Stochastic MPC by Certainty Equivalence

Let us mention that while in this section we have focused on deterministic problems, there are variants of MPC, which include the treatment of uncertainty. The books and papers cited earlier contain several ideas along these lines; see for example the books by Kouvaritakis and Cannon [KoC16], Rawlings, Mayne, and Diehl [RMD17], and the survey by Mesbah [Mes16].

In this connection it is also worth mentioning the *certainty equivalence approach* that we discussed briefly in Section 3.2. As noted in that section, upon reaching state x_k we may perform the MPC calculations after replacing the uncertain quantities $w_{k+1}, w_{k+2}, \ldots$ with deterministic quantities $\overline{w}_{k+1}, \overline{w}_{k+2}, \ldots$, *while allowing for the stochastic character of the disturbance w_k of just the current stage k*. This MPC calculation is not much more difficult that the one for deterministic problems, while still implementing a Newton step for solving the associated Bellman equation; see the discussion of Section 3.2, and also Section 2.5.3 of the RL book [Ber19a].

State Constraints, Target Tubes, and Off-Line Training

Our discussion so far has skirted a major issue in MPC, which is that there may be additional state constraints of the form $x_k \in X$, for all k, where X is some subset of the true state space. Indeed much of the original work on MPC was motivated by control problems with state constraints, imposed by the physics of the problem, which could not be handled effectively with the nice unconstrained framework of the linear quadratic problem that we have discussed in Chapter 4.

The treatment of state constraints is connected to the theory of reachability of target tubes, first formulated and studied by the author in his Ph.D. thesis [Ber71], and subsequent papers [BeR71], [Ber72]; see the books [Ber17a], [Ber19a], [Ber20a] for a discussion that is consistent with the viewpoint of this section. A target tube is a subset $\tilde{X}$ of the state constraint set X, within which the state can be kept indefinitely with feasible control choices, assuming that the initial state belongs to $\tilde{X}$. In other words, the problem (5.10) may not be feasible for every $x_k \in X$, once the constraint $x_t \in X$ for all $t = k+1, k+2, \ldots$, is added in problem (5.10). However, a suitable target tube is one specified by a subset $\tilde{X} \subset X$ such that the problem (5.10) is feasible under the constraint $x_t \in \tilde{X}$ for all $t = k+1, k+2, \ldots$, provided $x_k \in \tilde{X}$.

There are several ways to compute sets $\tilde{X}$ with this property, for which we refer to the aforementioned author's work and the MPC literature; see e.g., the book by Rawlings, Mayne, and Diehl [RMD17], and the survey by Mayne [May14]. The important point here is that the computation of a target tube must be done off-line with one of several available algorithmic approaches, so it becomes part of the off-line training (in addition to the terminal cost function G).

Given an off-line training process, which provides a target tube constraint $x_k \in \tilde{X}$ for all k, a terminal cost function G, and possibly one or more base policies for truncated rollout, MPC becomes an on-line play algorithm for which our earlier discussion applies. Note, however, that in an indirect adaptive control context, where a model is estimated on-line as it is changing, it may be difficult to recompute on-line a target tube that can be used to enforce the state constraints of the problem, particularly if the states constraints change themselves as part of the changing problem data. This is a problem-dependent issue that deserves further attention.

5.5 NOTES AND SOURCES

The literature for PID control is extensive and includes the books by Aström and Hagglund [AsH95], [AsH06]. For detailed accounts of adaptive control, we refer to the books by Aström and Wittenmark [AsW08], Bodson [Bod20], Goodwin and Sin [GoS84], Ioannou and Sun [IoS96], Jiang and Jiang [JiJ17], Krstic, Kanellakopoulos, and Kokotovic [KKK95], Kokotovic [Kok91], Kumar and Varaiya [KuV86], Liu, et al. [LWW17], Lavretsky and Wise [LaW13], Narendra and Annaswamy [NaA12], Sastry and Bodson [SaB11], Slotine and Li [SlL91], and Vrabie, Vamvoudakis, and Lewis [VVL13].

The literature on MPC is voluminous, and has grown over time to include problem and algorithm variations and extensions. For detailed accounts, we refer to the textbooks by Maciejowski [Mac02], Goodwin, Seron, and De Dona [GSD06], Camacho and Bordons [CaB07], Kouvaritakis and Cannon [KoC16], Borrelli, Bemporad, and Morari [BBM17], and Rawlings, Mayne, and Diehl [RMD17].

Deterministic optimal control with infinite state and control spaces can exhibit unusual/pathological behavior. For the case of nonnegative cost per stage, an analysis of the exact value and policy iteration algorithms, including convergence issues and counterexamples, is given in the author's paper [Ber17b] and abstract DP book [Ber22a]. The case of nonpositive cost per stage has been addressed in classical analyses, beginning with the work of Blackwell [Bla65]; see also [Str66], [BeS78], [YuB15].

6

Finite Horizon Deterministic Problems - Discrete Optimization

Contents
6.1. Deterministic Discrete Spaces Finite Horizon Problems . p. 120
6.2. General Discrete Optimization Problems . . . . p. 125
6.3. Approximation in Value Space . . . . p. 128
6.4. Rollout Algorithms for Discrete Optimization . . . . . p. 132
6.5. Rollout and Approximation in Value Space with Multistep Lookahead . . . . p. 149
6.5.1. Simplified Multistep Rollout - Double Rollout . . . . p. 150
6.5.2. Incremental Rollout for Multistep Approximation in Value Space . . . . p. 153
6.6. Constrained Forms of Rollout Algorithms . . . . p. 159
6.7. Adaptive Control by Rollout with a POMDP Formulation p. 173
6.8. Rollout for Minimax Control . . . . p. 181
6.9. Small Stage Costs and Long Horizon - Continuous-Time Rollout . . . . p. 190
6.10. Epilogue . . . . p. 197

120 *Finite Horizon Deterministic Problems - Discrete Optimization* *Chap. 6*

In this chapter, we discuss finite horizon deterministic problems, focusing primarily on the case where the state and control spaces are finite. After we introduce these problems, we will argue that they can be transformed to infinite horizon SSP problems, through the use of an artificial cost-free termination state that the system moves into at the end of the horizon. Once the problem is transformed to an infinite horizon SSP problem, the ideas of approximation in value space, off-line training, on-line play, and Newton's method, which we have developed earlier, become applicable. Moreover the ideas of MPC and adaptive control are easily adapted within the finite horizon discrete optimization framework.

An interesting aspect of our methodology for discrete deterministic problems is that it admits extensions that we have not discussed so far. The extensions include variants that apply to constrained forms of DP, which involve constraints on the entire system trajectory, and also allow the use of heuristic algorithms that are more general than policies within the context of rollout. These variants rely on the problem's deterministic structure, and do not extend to stochastic problems.

Another interesting aspect of discrete deterministic problems is that they can serve as a framework for an important class of commonly encountered discrete optimization problems, including integer programming and combinatorial optimization problems such as scheduling, assignment, routing, etc. This will bring to bear the methodology of approximation in value space, rollout, adaptive control, and MPC, and provide effective suboptimal solution methods for these problems.

In the present chapter, we provide a brief summary of approximation in value space and rollout algorithms, aimed to make the connection with approximation in value space and Newton's method for infinite horizon problems. Additional discussion may be found in the author's rollout and policy iteration book [Ber20a], on which this chapter is based. Moreover, in Section 6.7-6.9, we will discuss DP problems that have a methodological connection to deterministic finite horizon DP, but require various algorithmic extensions.

6.1 DETERMINISTIC DISCRETE SPACES FINITE HORIZON PROBLEMS

In deterministic finite horizon DP problems, the state is generated nonrandomly over N stages, and involves a system of the form

$$x_{k+1} = f_k(x_k, u_k), \qquad k = 0, 1, \ldots, N-1, \qquad (6.1)$$

where k is the time index, and

- x_k is the state of the system, an element of some state space X_k,
- u_k is the control or decision variable, to be selected at time k from some given set $U_k(x_k)$, a subset of a control space U_k, that depends on x_k,

Sec. 6.1 Deterministic Discrete Spaces Finite Horizon Problems

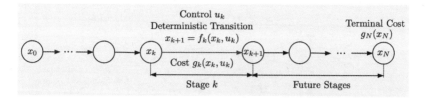

Figure 6.1.1 Illustration of a deterministic N-stage optimal control problem. Starting from state x_k, the next state under control u_k is generated nonrandomly, according to

$$x_{k+1} = f_k(x_k, u_k),$$

and a stage cost $g_k(x_k, u_k)$ is incurred.

f_k is a function of (x_k, u_k) that describes the mechanism by which the state is updated from time k to time $k+1$.

The state space X_k and control space U_k are arbitrary sets and may depend on k. Similarly the system function f_k can be arbitrary and may depend on k. The cost incurred at time k is denoted by $g_k(x_k, u_k)$, and the function g_k may depend on k. For a given initial state x_0, the total cost of a control sequence $\{u_0, \ldots, u_{N-1}\}$ is

$$J(x_0; u_0, \ldots, u_{N-1}) = g_N(x_N) + \sum_{k=0}^{N-1} g_k(x_k, u_k), \tag{6.2}$$

where $g_N(x_N)$ is a terminal cost incurred at the end of the process. This is a well-defined number, since the control sequence $\{u_0, \ldots, u_{N-1}\}$ together with x_0 determines exactly the state sequence $\{x_1, \ldots, x_N\}$ via the system equation (6.1); see Figure 6.1.1. We want to minimize the cost (6.2) over all sequences $\{u_0, \ldots, u_{N-1}\}$ that satisfy the control constraints, thereby obtaining the optimal value as a function of x_0

$$J^*(x_0) = \min_{\substack{u_k \in U_k(x_k) \\ k=0,\ldots,N-1}} J(x_0; u_0, \ldots, u_{N-1}).$$

Notice an important difference from the stochastic case: we optimize over sequences of controls $\{u_0, \ldots, u_{N-1}\}$, rather than over policies that consist of a sequence of functions $\pi = \{\mu_0, \ldots, \mu_{N-1}\}$, where μ_k maps states x_k into controls $u_k = \mu_k(x_k)$, and satisfies the control constraints $\mu_k(x_k) \in U_k(x_k)$ for all x_k. It is well-known that in the presence of stochastic uncertainty, policies are more effective than control sequences, and can result in improved cost. On the other hand for deterministic problems, minimizing over control sequences yields the same optimal cost as over policies, since the cost of any policy starting from a given state determines with certainty the controls applied at that state and the future states, and hence can also be achieved by the corresponding control sequence. This

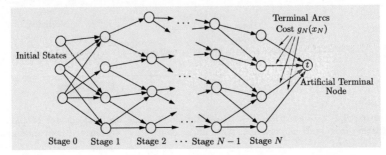

Figure 6.1.2 Transition graph for a deterministic system with finite control space and a finite number of initial states. Nodes correspond to states x_k. Arcs correspond to state-control pairs (x_k, u_k). An arc (x_k, u_k) has start and end nodes x_k and $x_{k+1} = f_k(x_k, u_k)$, respectively. The transition cost $g_k(x_k, u_k)$ is viewed as the length of this arc. The problem is equivalent to finding a shortest path from initial nodes of stage 0 to an artificial terminal node t.

point of view allows more general forms of rollout, which we will discuss in this section: instead of using a policy for rollout, we will allow the use of more general heuristics for choosing future controls.

Discrete Optimal Control - Transformation to an Infinite Horizon Problem

We use the term *discrete optimal control* to refer to deterministic DP problems where the control spaces are either naturally discrete and consist of a finite number of elements, or have been discretized for the purposes of computation. Generally, whenever we assume that the control space is finite, we will also assume implicitly a single or at most a finite number of possible initial states, so the number of states that can be generated at each stage is also finite. A problem of this type can be conveniently described with an acyclic graph specifying for each state x_k the possible transitions to next states x_{k+1}. The nodes of the graph correspond to states x_k and the arcs of the graph correspond to state-control pairs (x_k, u_k). Each arc with start node x_k corresponds to a choice of a single control $u_k \in U_k(x_k)$ and has as end node the next state $f_k(x_k, u_k)$. The cost of an arc (x_k, u_k) is defined as $g_k(x_k, u_k)$; see Fig. 6.1.2. To handle the final stage, an artificial terminal node t is added. Each state x_N at stage N is connected to the terminal node t with an arc having cost $g_N(x_N)$.

Note that control sequences $\{u_0, \ldots, u_{N-1}\}$ correspond to paths originating at the initial state (a node at stage 0) and terminating at one of the nodes corresponding to the final stage N. If we view the cost of an arc as its length, we see that *a deterministic finite-state finite horizon problem is equivalent to finding a minimum-length (or shortest) path from the initial nodes of the graph (stage 0) to the terminal node t*. Here, by the length of a path we mean the sum of the lengths of its arcs. It also turns out that the reverse is true: every shortest path problem involving a graph whose

Sec. 6.1   Deterministic Discrete Spaces Finite Horizon Problems   123

cycles have positive length can be transformed into a discrete optimal control problem. This fact is important, but will not be useful to us, so we will not consider it further here (see the textbook [Ber17a], Chapter 2, for a detailed discussion).

The connection of finite state and control spaces finite horizon deterministic problem with a shortest path problem is important for our purposes. The reason is that it provides a bridge to an SSP problem with an infinite horizon, and by extension, to our earlier development of approximation in value space, Newton's method, rollout, and the PI algorithm. It is also important to recognize that this SSP problem has a few additional special characteristics. These are:

(a) The equivalent SSP involves a deterministic system, has a finite number of states and controls, and involves an acyclic graph. The states of the SSP are all the state-time pairs (x_k, k), $k = 0, 1, \ldots, N$, where x_k is one of the finite number of elements of X_k that are reachable from one of the finite number of initial states x_0 using a feasible sequence of controls. The possible transitions from states (x_k, k) to states $(x_{k+1}, k+1)$ correspond to controls $u_k \in U_k(x_k)$ such that $x_{k+1} = f_k(x_k, u_k)$.

(b) The state space of the SSP expands as the horizon N becomes longer. While this complicates the use of the PI algorithm, it does not materially affect the use of a rollout algorithm.

(c) The optimal cost function of the SSP is obtained from the optimal cost functions of the finite horizon problem, which are generated by the DP algorithm to be presented shortly. This DP algorithm can be viewed as the Bellman equation of the SSP.

The Exact Dynamic Programming Algorithm

The DP algorithm for finite horizon deterministic problems rests on a simple idea, the *principle of optimality*, which suggests that the optimal cost function can be constructed in piecemeal fashion going backwards: first compute the optimal cost function for the "tail subproblem" involving the last stage, then solve the "tail subproblem" involving the last two stages, and continue in this manner until the optimal cost function for the entire problem is constructed.

By translating into mathematical terms the principle of optimality, we obtain the DP algorithm. It constructs functions

$$J_N^*(x_N), J_{N-1}^*(x_{N-1}), \ldots, J_0^*(x_0),$$

sequentially, starting from J_N^*, and proceeding backwards to J_{N-1}^*, J_{N-2}^*, etc. The value $J_k^*(x_k)$ will be viewed as the optimal cost of the tail subproblem that starts at state x_k at time k and ends at a state x_N.

DP Algorithm for Deterministic Finite Horizon Problems

Start with
$$J_N^*(x_N) = g_N(x_N), \qquad \text{for all } x_N, \tag{6.3}$$
and for $k = 0, \ldots, N-1$, let
$$J_k^*(x_k) = \min_{u_k \in U_k(x_k)} \Big[g_k(x_k, u_k) + J_{k+1}^*\big(f_k(x_k, u_k)\big) \Big], \qquad \text{for all } x_k. \tag{6.4}$$

Note that at stage k, the calculation in Eq. (6.4) must be done for all states x_k before proceeding to stage $k-1$. The key fact about the DP algorithm is that for every initial state x_0, the number $J_0^*(x_0)$ obtained at the last step, is equal to the optimal cost $J^*(x_0)$. Indeed, a more general fact can be shown, namely that for all $k = 0, 1, \ldots, N-1$, and all states x_k at time k, we have

$$J_k^*(x_k) = \min_{\substack{u_m \in U_m(x_m) \\ m = k, \ldots, N-1}} J(x_k; u_k, \ldots, u_{N-1}), \tag{6.5}$$

where $J(x_k; u_k, \ldots, u_{N-1})$ is the cost generated by starting at x_k and using subsequent controls $u_k, \ldots, u_{N-1}$:

$$J(x_k; u_k, \ldots, u_{N-1}) = g_N(x_N) + \sum_{t=k}^{N-1} g_t(x_t, u_t).$$

Thus, $J_k^*(x_k)$ is the optimal cost for an $(N-k)$-stage tail subproblem that starts at state x_k and time k, and ends at time N. Based on this interpretation of $J_k^*(x_k)$, we call it the *optimal cost-to-go* from state x_k at stage k, and refer to J_k^* as the *optimal cost-to-go function* or *optimal cost function* at time k.

Once the functions $J_0^*, \ldots, J_N^*$ have been obtained, we can use a forward algorithm to construct an optimal control sequence $\{u_0^*, \ldots, u_{N-1}^*\}$ and state trajectory $\{x_1^*, \ldots, x_N^*\}$ for a given initial state x_0.

Construction of Optimal Control Sequence $\{u_0^*, \ldots, u_{N-1}^*\}$

Set
$$u_0^* \in \arg\min_{u_0 \in U_0(x_0)} \Big[g_0(x_0, u_0) + J_1^*\big(f_0(x_0, u_0)\big) \Big],$$
and
$$x_1^* = f_0(x_0, u_0^*).$$

Sec. 6.2   General Discrete Optimization Problems   125

Sequentially, going forward, for $k = 1, 2, \ldots, N-1$, set

$$u_k^* \in \arg\min_{u_k \in U_k(x_k^*)} \left[g_k(x_k^*, u_k) + J_{k+1}^*\bigl(f_k(x_k^*, u_k)\bigr)\right], \qquad (6.6)$$

and

$$x_{k+1}^* = f_k(x_k^*, u_k^*).$$

Note an interesting conceptual division of the optimal control sequence construction: there is "off-line training" to obtain J_k^* by precomputation [cf. the DP Eqs. (6.3)-(6.4)], which is followed by "on-line play" in real-time to obtain u_k^* [cf. Eq. (6.6)]. This is analogous to the two algorithmic processes described in Chapter 1 in connection with chess and backgammon.

6.2 GENERAL DISCRETE OPTIMIZATION PROBLEMS

Discrete deterministic optimization problems, including challenging combinatorial problems, can be typically formulated as DP problems by breaking down each feasible solution into a sequence of decisions/controls. This formulation often leads to an intractable exact DP computation because of an exponential explosion of the number of states as time progresses. However, a reformulation to a discrete optimal control brings to bear approximate DP methods, such as rollout and others, to be discussed shortly, which can deal with the exponentially increasing size of the state space. We illustrate the reformulation with an example and then generalize.

Example 6.2.1 (The Traveling Salesman Problem)

An important model for scheduling a sequence of operations is the classical traveling salesman problem. Here we are given N cities and the travel time between each pair of cities. We wish to find a minimum time travel that visits each of the cities exactly once and returns to the start city. To convert this problem to a DP problem, we form a graph whose nodes are the sequences of k distinct cities, where $k = 1, \ldots, N$. The k-city sequences correspond to the states of the kth stage. The initial state x_0 consists of some city, taken as the start (city A in the example of Fig. 6.2.1). A k-city node/state leads to a $(k+1)$-city node/state by adding a new city at a cost equal to the travel time between the last two of the $k+1$ cities; see Fig. 6.2.1. Each sequence of N cities is connected to an artificial terminal node t with an arc of cost equal to the travel time from the last city of the sequence to the starting city, thus completing the transformation to a DP problem.

The optimal costs-to-go from each node to the terminal state can be obtained by the DP algorithm and are shown next to the nodes. Note, how-

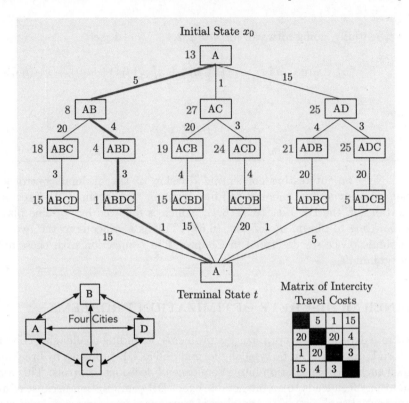

Figure 6.2.1 Example of a DP formulation of the traveling salesman problem. The travel times between the four cities A, B, C, and D are shown in the matrix at the bottom. We form a graph whose nodes are the k-city sequences and correspond to the states of the kth stage, assuming that A is the starting city. The transition costs/travel times are shown next to the arcs. The optimal costs-to-go are generated by DP starting from the terminal state and going backwards towards the initial state, and are shown next to the nodes. There is a unique optimal sequence here (ABDCA), and it is marked with thick lines. The optimal sequence can be obtained by forward minimization [cf. Eq. (6.6)], starting from the initial state x_0.

ever, that the number of nodes grows exponentially with the number of cities N. This makes the DP solution intractable for large N. As a result, large traveling salesman and related scheduling problems are typically addressed with approximation methods, some of which are based on DP, and will be discussed later.

Let us now extend the ideas of the preceding example to the general discrete optimization problem:

$$\text{minimize } G(u)$$
$$\text{subject to } u \in U,$$

Sec. 6.2   General Discrete Optimization Problems   127

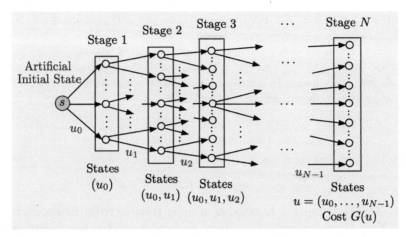

Figure 6.2.2 Formulation of a discrete optimization problem as a DP problem with N stages. There is a cost $G(u)$ only at the terminal stage on the arc connecting an N-solution $u = (u_0, \ldots, u_{N-1})$ upon reaching the terminal state. Note that there is only one incoming arc at each node.

where U is a finite set of feasible solutions and $G(u)$ is a cost function. We assume that each solution u has N components; i.e., it has the form $u = (u_0, \ldots, u_{N-1})$, where N is a positive integer. We can then view the problem as a sequential decision problem, where the components $u_0, \ldots, u_{N-1}$ are selected one-at-a-time. A k-tuple $(u_0, \ldots, u_{k-1})$ consisting of the first k components of a solution is called a k-*solution*. We associate k-solutions with the kth stage of the finite horizon discrete optimal control problem shown in Fig. 6.2.2. In particular, for $k = 1, \ldots, N$, we view as the states of the kth stage all the k-tuples $(u_0, \ldots, u_{k-1})$. For stage $k = 0, \ldots, N-1$, we view u_k as the control. The initial state is an artificial state denoted s. From this state, by applying u_0, we may move to any "state" (u_0), with u_0 belonging to the set

$$U_0 = \{\tilde{u}_0 \mid \text{there exists a solution of the form } (\tilde{u}_0, \tilde{u}_1, \ldots, \tilde{u}_{N-1}) \in U\}. \tag{6.7}$$

Thus U_0 is the set of choices of u_0 that are consistent with feasibility.

More generally, from a state $(u_0, \ldots, u_{k-1})$, we may move to any state of the form $(u_0, \ldots, u_{k-1}, u_k)$, upon choosing a control u_k that belongs to the set

$$U_k(u_0, \ldots, u_{k-1}) = \{u_k \mid \text{for some } \overline{u}_{k+1}, \ldots, \overline{u}_{N-1} \text{ we have}$$
$$(u_0, \ldots, u_{k-1}, u_k, \overline{u}_{k+1}, \ldots, \overline{u}_{N-1}) \in U\}. \tag{6.8}$$

These are the choices of u_k that are consistent with the preceding choices $u_0, \ldots, u_{k-1}$, and are also consistent with feasibility. The last stage corresponds to the N-solutions $u = (u_0, \ldots, u_{N-1})$, and the terminal cost is

$G(u)$; see Fig. 6.2.2. All other transitions in this DP problem formulation have cost 0.

Let $J_k^*(u_0, \ldots, u_{k-1})$ denote the optimal cost starting from the k-solution $(u_0, \ldots, u_{k-1})$, i.e., the optimal cost of the problem over solutions whose first k components are constrained to be equal to $u_0, \ldots, u_{k-1}$. The DP algorithm is described by the equation

$$J_k^*(u_0, \ldots, u_{k-1}) = \min_{u_k \in U_k(u_0, \ldots, u_{k-1})} J_{k+1}^*(u_0, \ldots, u_{k-1}, u_k),$$

with the terminal condition

$$J_N^*(u_0, \ldots, u_{N-1}) = G(u_0, \ldots, u_{N-1}).$$

This algorithm executes backwards in time: starting with the known function $J_N^* = G$, we compute J_{N-1}^*, then J_{N-2}^*, and so on up to computing J_0^*. An optimal solution $(u_0^*, \ldots, u_{N-1}^*)$ is then constructed by going forward through the algorithm

$$u_k^* \in \arg \min_{u_k \in U_k(u_0^*, \ldots, u_{k-1}^*)} J_{k+1}^*(u_0^*, \ldots, u_{k-1}^*, u_k), \quad k = 0, \ldots, N-1, \quad (6.9)$$

where U_0 is given by Eq. (6.7), and U_k is given by Eq. (6.8): first compute u_0^*, then u_1^*, and so on up to u_{N-1}^*; cf. Eq. (6.6).

Of course here the number of states typically grows exponentially with N, but we can use the DP minimization (6.9) as a starting point for approximation methods. For example we may try to use approximation in value space, whereby we replace J_{k+1}^* with some suboptimal $\tilde{J}_{k+1}$ in Eq. (6.9). One possibility is to use as

$$\tilde{J}_{k+1}(u_0^*, \ldots, u_{k-1}^*, u_k),$$

the cost generated by a heuristic method that solves the problem suboptimally with the values of the first $k+1$ decision components fixed at $u_0^*, \ldots, u_{k-1}^*, u_k$. This is the *rollout algorithm*, which turns out to be a very simple and effective approach for approximate combinatorial optimization.

Let us finally note that while we have used a general cost function G and constraint set U in our discrete optimization model of this section, in many problems G and/or U may have a special (e.g., additive) structure, which is consistent with a sequential decision making process and may be computationally exploited. The traveling salesman Example 6.2.1 is a case in point, where G consists of N components (the intercity travel costs), one per stage.

6.3 APPROXIMATION IN VALUE SPACE

The forward optimal control sequence construction of Eq. (6.6) is possible only after we have computed $J_k^*(x_k)$ by DP for all x_k and k. Unfortunately,

Sec. 6.3   Approximation in Value Space

in practice this is often prohibitively time-consuming, because the number of possible pairs (x_k, k) can be very large. However, a similar forward algorithmic process can be used if the optimal cost-to-go functions J_k^* are replaced by some approximations $\tilde{J}_k$. This is the idea of approximation in value space that we have discussed earlier in connection with infinite horizon problems. It constructs a suboptimal solution $\{\tilde{u}_0, \ldots, \tilde{u}_{N-1}\}$ in place of the optimal $\{u_0^*, \ldots, u_{N-1}^*\}$, by using $\tilde{J}_k$ in place of J_k^* in the DP procedure (6.6). Based on our infinite horizon analysis of Chapter 3 and the interpretation of the deterministic finite horizon problem as an infinite horizon SSP, *the cost function of the corresponding one-step lookahead policy can be viewed as the result of a Newton step for solving Bellman's equation, i.e., the DP algorithm (6.4), starting from the point* $(\tilde{J}_1, \tilde{J}_2, \ldots, \tilde{J}_N)$.

Approximation in Value Space - Use of $\tilde{J}_k$ in Place of J_k^*

Start with

$$\tilde{u}_0 \in \arg \min_{u_0 \in U_0(x_0)} \Big[g_0(x_0, u_0) + \tilde{J}_1\big(f_0(x_0, u_0)\big)\Big],$$

and set

$$\tilde{x}_1 = f_0(x_0, \tilde{u}_0).$$

Sequentially, going forward, for $k = 1, 2, \ldots, N-1$, set

$$\tilde{u}_k \in \arg \min_{u_k \in U_k(\tilde{x}_k)} \Big[g_k(\tilde{x}_k, u_k) + \tilde{J}_{k+1}\big(f_k(\tilde{x}_k, u_k)\big)\Big], \qquad (6.10)$$

and

$$\tilde{x}_{k+1} = f_k(\tilde{x}_k, \tilde{u}_k).$$

Thus in approximation in value space the calculation of the suboptimal sequence $\{\tilde{u}_0, \ldots, \tilde{u}_{N-1}\}$ is done by going forward (no backward calculation is needed once the approximate cost-to-go functions $\tilde{J}_k$ are available). This is similar to the calculation of the optimal sequence $\{u_0^*, \ldots, u_{N-1}^*\}$ [cf. Eq. (6.6)], and is independent of how the functions $\tilde{J}_k$ are computed.

An alternative (and equivalent) form of the exact DP algorithm (6.4), uses the optimal cost-to-go functions J_k^* indirectly. In particular, it generates the *optimal Q-factors*, defined for all pairs (x_k, u_k) and k by

$$Q_k^*(x_k, u_k) = g_k(x_k, u_k) + J_{k+1}^*\big(f_k(x_k, u_k)\big). \qquad (6.11)$$

Thus the optimal Q-factors are simply the expressions that are minimized in the right-hand side of the DP equation (6.4).

Note that the optimal cost function J_k^* can be recovered from the optimal Q-factor Q_k^* by means of the minimization

$$J_k^*(x_k) = \min_{u_k \in U_k(x_k)} Q_k^*(x_k, u_k). \tag{6.12}$$

Moreover, the DP algorithm (6.4) can be written in an essentially equivalent form that involves Q-factors only [cf. Eqs. (6.11)-(6.12)]:

$$Q_k^*(x_k, u_k) = g_k(x_k, u_k) + \min_{u_{k+1} \in U_{k+1}(f_k(x_k, u_k))} Q_{k+1}^*\big(f_k(x_k, u_k), u_{k+1}\big).$$

Exact and approximate forms of this and other related algorithms, including counterparts for stochastic optimal control problems, comprise an important class of RL methods known as *Q-learning*.

The expression

$$\tilde{Q}_k(x_k, u_k) = g_k(x_k, u_k) + \tilde{J}_{k+1}\big(f_k(x_k, u_k)\big),$$

which is minimized in approximation in value space [cf. Eq. (6.10)] is known as the (approximate) *Q-factor of* (x_k, u_k). Note that the computation of the suboptimal control (6.10) can be done through the Q-factor minimization

$$\tilde{u}_k \in \arg\min_{u_k \in U_k(\tilde{x}_k)} \tilde{Q}_k(\tilde{x}_k, u_k).$$

This suggests the possibility of using approximate off-line trained Q-factors in place of cost functions in approximation in value space schemes. However, contrary to the cost approximation scheme (6.10) and its multistep counterparts, the performance may be degraded through the errors in the off-line training of the Q-factors (depending on how the training is done).

Multistep Lookahead

The approximation in value space algorithm (6.10) involves a one-step lookahead minimization, since it solves a one-stage DP problem for each k. We may also consider ℓ-*step lookahead*, which involves the solution of an ℓ-step DP problem, where ℓ is an integer, $1 < \ell < N - k$, with a terminal cost function approximation $\tilde{J}_{k+\ell}$. This is similar to the infinite horizon case that we discussed in Chapter 2. As we have noted in that section, multistep lookahead typically provides better performance over one-step lookahead in RL approximation schemes. For example in AlphaZero chess, long multistep lookahead is critical for good on-line performance. On the negative side, the solution of the multistep lookahead optimization problem, instead of the one-step lookahead counterpart of Eq. (6.10), becomes more time consuming.

Rollout

Similar to infinite horizon problems, a major issue in the value space approximation (6.10) is the construction of suitable approximate cost-to-go functions $\tilde{J}_{k+1}$. This can be done in many different ways, including some of the principal RL methods. For example, $\tilde{J}_{k+1}$ may be constructed with a sophisticated off-line training method, as discussed in Section 1.1, in connection with chess and backgammon. Forms of approximate PI method can be applied in particular, possibly with the use of neural networks, once the problem is viewed as an infinite horizon SSP problem. Another possibility is the fitted value iteration method, which is described in Section 4.3 of the book [Ber19a], and Section 4.3.1 of the book [Ber20a].

Alternatively, $\tilde{J}_{k+1}$ may be obtained on-line with *rollout*, whereby the approximate values $\tilde{J}_{k+1}(x_{k+1})$ are obtained when needed by running a heuristic control scheme, called *base heuristic*, for a suitably large number of steps, starting from x_{k+1}.† The base heuristic need not be a policy. It could be any method, which starting from a state x_{k+1} generates a sequence controls $u_{k+1}, \ldots, u_{N-1}$, the corresponding sequence of states $x_{k+2}, \ldots, x_N$, and the cost of the heuristic starting from x_{k+1}, which we will generically denote by $H_{k+1}(x_{k+1})$ in this chapter:

$$H_{k+1}(x_{k+1}) = g_{k+1}(x_{k+1}, u_{k+1}) + \cdots + g_{N-1}(x_{N-1}, u_{N-1}) + g_N(x_N).$$

This value of $H_{k+1}(x_{k+1})$ is the one used as the approximate cost-to-go $\tilde{J}_{k+1}(x_{k+1})$ in the corresponding approximation in value space scheme (6.10). An important point here is that deterministic problems hold a special attraction for rollout, as they do not require expensive on-line Monte Carlo simulation to calculate the cost function values $\tilde{J}_{k+1}(x_{k+1})$.

There are also several variants of rollout, involving for example truncation, multistep lookahead, and other possibilities. In particular, truncated rollout combines the use of one-step optimization, simulation of the

---

† For deterministic problems we prefer to use the term "base heuristic" rather than "base policy" for reasons to be explained later in this chapter, in the context of the notion of sequential consistency (the heuristic may not qualify as a legitimate DP policy). In particular, if the base heuristic, when started at state x_k, generates the sequence $\{\tilde{u}_k, \tilde{x}_{k+1}, \tilde{u}_{k+1}, \tilde{x}_{k+2}, \ldots, \tilde{u}_{N-1}, \tilde{x}_N\}$, it is not necessarily true that, when started at state $\tilde{x}_{k+1}$, it will generate the tail portion that starts at $\tilde{u}_{k+1}$, namely $\{\tilde{u}_{k+1}, \tilde{x}_{k+2}, \ldots, \tilde{u}_{N-1}, \tilde{x}_N\}$, (which would be true if the heuristic were a legitimate policy). More generally, the method used by the base heuristic to complete the system's trajectory starting from some state may be very different than the method used to complete the trajectory starting at another state. In any case, if the base heuristic is not a legitimate policy, then the use of $H_{k+1}(x_{k+1})$ as terminal cost function approximation, yields a type of approximation in value space scheme, which can still be interpreted as a Newton step.

132 *Finite Horizon Deterministic Problems - Discrete Optimization* Chap. 6

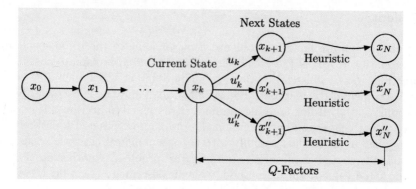

Figure 6.4.1 Schematic illustration of rollout with one-step lookahead for a deterministic problem. At state x_k, for every pair (x_k, u_k), $u_k \in U_k(x_k)$, the base heuristic generates a Q-factor

$$\tilde{Q}_k(x_k, u_k) = g_k(x_k, u_k) + H_{k+1}\big(f_k(x_k, u_k)\big),$$

and the rollout algorithm selects the control $\tilde{\mu}_k(x_k)$ with minimal Q-factor.

base policy for a certain number of steps m, and then adds an approximate cost $\tilde{J}_{k+m+1}(x_{k+m+1})$ to the cost of the simulation, which depends on the state x_{k+m+1} obtained at the end of the rollout. Note that if one foregoes the use of a base heuristic (i.e., $m = 0$), one recovers as a special case the general approximation in value space scheme. Versions of truncated rollout with multistep lookahead minimization are also possible. Other variants of rollout include versions involving multiple heuristics, combinations with other forms of approximation in value space methods, and multistep lookahead, which will be described later in this chapter, starting with the next section.

6.4 ROLLOUT ALGORITHMS FOR DISCRETE OPTIMIZATION

In this section, we will develop in more detail the theory of rollout with one-step lookahead minimization for deterministic problems, including the important issue of cost improvement. We will also illustrate several variants of the method, and we will consider questions of efficient implementation. We will then discuss examples of discrete optimization applications.

Let us consider a deterministic DP problem with a finite number of controls and a given initial state (so the number of states that can be reached from the initial state is also finite). We first focus on the pure form of rollout that uses one-step lookahead and no terminal cost approximation. Given a state x_k at time k, this algorithm considers the tail subproblems that start at every possible next state x_{k+1}, and solves them suboptimally by using some algorithm, referred to as *base heuristic*.

Sec. 6.4 Rollout Algorithms for Discrete Optimization

Thus when at x_k, rollout generates on-line the next states x_{k+1} that correspond to all $u_k \in U_k(x_k)$, and uses the base heuristic to compute the sequence of states $\{x_{k+1}, \ldots, x_N\}$ and controls $\{u_{k+1}, \ldots, u_{N-1}\}$ such that

$$x_{t+1} = f_t(x_t, u_t), \qquad t = k, \ldots, N-1,$$

and the corresponding cost

$$H_{k+1}(x_{k+1}) = g_{k+1}(x_{k+1}, u_{k+1}) + \cdots + g_{N-1}(x_{N-1}, u_{N-1}) + g_N(x_N).$$

The rollout algorithm then applies the control that minimizes over $u_k \in U_k(x_k)$ the tail cost expression for stages k to N:

$$g_k(x_k, u_k) + H_{k+1}(x_{k+1}).$$

Equivalently, and more succinctly, the rollout algorithm applies at state x_k the control $\tilde{\mu}_k(x_k)$ given by the minimization

$$\tilde{\mu}_k(x_k) \in \arg\min_{u_k \in U_k(x_k)} \tilde{Q}_k(x_k, u_k), \tag{6.13}$$

where $\tilde{Q}_k(x_k, u_k)$ is the approximate Q-factor defined by

$$\tilde{Q}_k(x_k, u_k) = g_k(x_k, u_k) + H_{k+1}\big(f_k(x_k, u_k)\big); \tag{6.14}$$

see Fig. 6.4.1. The rollout algorithm thus defines a suboptimal policy $\tilde{\pi} = \{\tilde{\mu}_0, \ldots, \tilde{\mu}_{N-1}\}$, referred to as the *rollout policy*, where for each x_k and k, $\tilde{\mu}_k(x_k)$ is the control produced by the Q-factor minimization (6.13).

Note that the rollout algorithm requires running the base heuristic for a number of times that is bounded by Nn, where n is an upper bound on the number of control choices available at each state. Thus if n is small relative to N, it requires computation equal to a small multiple of N times the computation time for a single application of the base heuristic. Similarly, if n is bounded by a polynomial in N, the ratio of the rollout algorithm computation time to the base heuristic computation time is a polynomial in N.

Example 6.4.1 (Traveling Salesman Problem)

Let us consider the traveling salesman problem of Example 6.2.1, whereby a salesman wants to find a minimum cost tour that visits each of N given cities $c = 0, \ldots, N-1$ exactly once and returns to the city he started from. With each pair of distinct cities c, c', we associate a traversal cost $g(c, c')$. Note that we assume that we can go directly from every city to every other city. There is no loss of generality in doing so because we can assign a very high cost $g(c, c')$ to any pair of cities (c, c') that is precluded from participation in

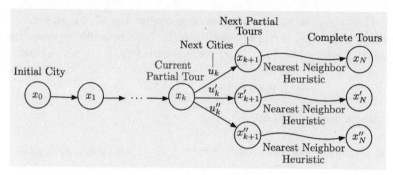

Figure 6.4.2 Rollout with the nearest neighbor heuristic for the traveling salesman problem. The initial state x_0 consists of a single city. The final state x_N is a complete tour of N cities, containing each city exactly once.

the solution. The problem is to find a visit order that goes through each city exactly once and whose sum of costs is minimum.

There are many heuristic approaches for solving the traveling salesman problem. For illustration purposes, let us focus on the simple *nearest neighbor* heuristic, which starts with a partial tour, i.e., an ordered collection of distinct cities, and constructs a sequence of partial tours, adding to the each partial tour a new city that does not close a cycle and minimizes the cost of the enlargement. In particular, given a sequence $\{c_0, c_1, \ldots, c_k\}$ consisting of distinct cities, the nearest neighbor heuristic adds a city c_{k+1} that minimizes $g(c_k, c_{k+1})$ over all cities $c_{k+1} \neq c_0, \ldots, c_k$, thereby forming the sequence $\{c_0, c_1, \ldots, c_k, c_{k+1}\}$. Continuing in this manner, the heuristic eventually forms a sequence of N cities, $\{c_0, c_1, \ldots, c_{N-1}\}$, thus yielding a complete tour with cost

$$g(c_0, c_1) + \cdots + g(c_{N-2}, c_{N-1}) + g(c_{N-1}, c_0). \qquad (6.15)$$

We can formulate the traveling salesman problem as a DP problem as we discussed in Example 6.2.1. We choose a starting city, say c_0, as the initial state x_0. Each state x_k corresponds to a partial tour $(c_0, c_1, \ldots, c_k)$ consisting of distinct cities. The states x_{k+1}, next to x_k, are sequences of the form $(c_0, c_1, \ldots, c_k, c_{k+1})$ that correspond to adding one more unvisited city $c_{k+1} \neq c_0, c_1, \ldots, c_k$ (thus the unvisited cities are the feasible controls at a given partial tour/state). The terminal states x_N are the complete tours of the form $(c_0, c_1, \ldots, c_{N-1}, c_0)$, and the cost of the corresponding sequence of city choices is the cost of the corresponding complete tour given by Eq. (6.15). Note that the number of states at stage k increases exponentially with k, and so does the computation required to solve the problem by exact DP.

Let us now use as a base heuristic the nearest neighbor method. The corresponding rollout algorithm operates as follows: After $k < N - 1$ iterations, we have a state x_k, i.e., a sequence $\{c_0, \ldots, c_k\}$ consisting of distinct cities. At the next iteration, we add one more city by running the nearest neighbor heuristic starting from each of the sequences of the form $\{c_0, \ldots, c_k, c\}$ where $c \neq c_0, \ldots, c_k$. We then select as next city c_{k+1} the city c that yielded the minimum cost tour under the nearest neighbor heuristic;

Sec. 6.4   *Rollout Algorithms for Discrete Optimization*   **135**

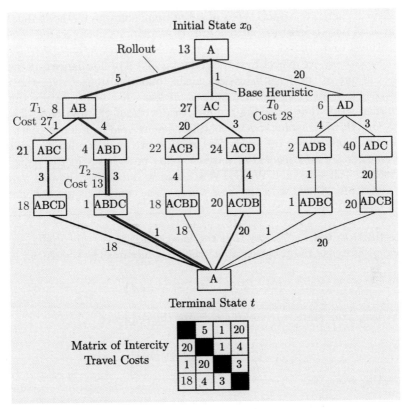

Figure 6.4.3 A traveling salesman problem example of rollout with the nearest neighbor base heuristic. At city A, the nearest neighbor heuristic generates the tour ACDBA (labelled T_0). At city A, the rollout algorithm compares the tours ABCDA, ACDBA, and ADCBA, finds ABCDA (labelled T_1) to have the least cost, and moves to city B. At AB, the rollout algorithm compares the tours ABCDA and ABDCA, finds ABDCA (labelled T_2) to have the least cost, and moves to city D. The rollout algorithm then moves to cities C and A (it has no other choice). Note that the algorithm generates three tours/solutions, T_0, T_1, and T_2, of decreasing costs 28, 27, and 13, respectively. The first path T_0 is generated by the base heuristic starting from the initial state, while the last tour T_2 is generated by rollout. This is suggestive of a general result that we will prove later: the rollout solution is better in terms of cost than the base heuristic solution. In fact the tour T_2 generated by rollout is optimal, but this is just a coincidence.

see Fig. 6.4.2. The overall computation for the rollout solution is bounded by a polynomial in N, and is much smaller than the exact DP computation. Figure 6.4.3 provides an example where the nearest neighbor heuristic and the corresponding rollout algorithm are compared.

Cost Improvement with Rollout - Sequential Consistency

The definition of the rollout algorithm leaves open the choice of the base heuristic. There are several types of suboptimal solution methods that can be used as base heuristics, such as greedy algorithms, local search, genetic algorithms, and others.

Intuitively, we expect that the rollout policy's performance is no worse than the one of the base heuristic: since rollout optimizes over the first control before applying the heuristic, it makes sense to conjecture that it performs better than applying the heuristic without the first control optimization. However, some special conditions must hold in order to guarantee this cost improvement property. We provide two such conditions, *sequential consistency* and *sequential improvement*, introduced in the paper by Bertsekas, Tsitsiklis, and Wu [BTW97], and we later show how to modify the algorithm to deal with the case where these conditions are not met.

Definition 6.4.1: We say that the base heuristic is *sequentially consistent* if it has the property that when it generates the sequence

$$\{x_k, u_k, x_{k+1}, u_{k+1}, \ldots, x_N\}$$

starting from state x_k, it also generates the sequence

$$\{x_{k+1}, u_{k+1}, \ldots, x_N\}$$

starting from state x_{k+1}.

In other words, the base heuristic is sequentially consistent if it "stays the course": when the starting state x_k is moved forward to the next state x_{k+1} of its state trajectory, the heuristic will not deviate from the remainder of the trajectory.

As an example, the reader may verify that the nearest neighbor heuristic described in the traveling salesman Example 6.4.1 is sequentially consistent. Similar examples include the use of many types of greedy/myopic heuristics (Section 6.4 of the book [Ber17a] provides some additional examples). Generally most heuristics used in practice satisfy the sequential consistency condition at "most" states x_k. However, some heuristics of interest may violate this condition at some states.

A sequentially consistent base heuristic can be recognized by the fact that it will apply the same control u_k at a state x_k, no matter what position x_k occupies in a trajectory generated by the base heuristic. Thus *a base heuristic is sequentially consistent if and only if it defines a legitimate DP policy*. This is the policy that moves from x_k to the state x_{k+1} that lies on the state trajectory $\{x_k, x_{k+1}, \ldots, x_N\}$ that the base heuristic generates.

We will now show that the rollout algorithm obtained with a sequentially consistent base heuristic yields no worse cost than the base heuristic.

Sec. 6.4   Rollout Algorithms for Discrete Optimization   137

> **Proposition 6.4.1: (Cost Improvement Under Sequential Consistency)** Consider the rollout policy $\tilde{\pi} = \{\tilde{\mu}_0, \ldots, \tilde{\mu}_{N-1}\}$ obtained with a sequentially consistent base heuristic, and let $J_{k,\tilde{\pi}}(x_k)$ denote the cost obtained with $\tilde{\pi}$ starting from x_k at time k. Then
>
> $$J_{k,\tilde{\pi}}(x_k) \leq H_k(x_k), \qquad \text{for all } x_k \text{ and } k, \tag{6.16}$$
>
> where $H_k(x_k)$ denotes the cost of the base heuristic starting from x_k.

Proof: We prove this inequality by induction. Clearly it holds for $k = N$, since
$$J_{N,\tilde{\pi}} = H_N = g_N.$$
Assume that it holds for index $k+1$. For any state x_k, let $\overline{u}_k$ be the control applied by the base heuristic at x_k. Then we have

$$\begin{aligned}
J_{k,\tilde{\pi}}(x_k) &= g_k\big(x_k, \tilde{\mu}_k(x_k)\big) + J_{k+1,\tilde{\pi}}\Big(f_k\big(x_k, \tilde{\mu}_k(x_k)\big)\Big) \\
&\leq g_k\big(x_k, \tilde{\mu}_k(x_k)\big) + H_{k+1}\Big(f_k\big(x_k, \tilde{\mu}_k(x_k)\big)\Big) \\
&= \min_{u_k \in U_k(x_k)} \Big[g_k(x_k, u_k) + H_{k+1}\big(f_k(x_k, u_k)\big) \Big] \\
&\leq g_k(x_k, \overline{u}_k) + H_{k+1}\big(f_k(x_k, \overline{u}_k)\big) \\
&= H_k(x_k),
\end{aligned} \tag{6.17}$$

where:

(a) The first equality is the DP equation for the rollout policy $\tilde{\pi}$.

(b) The first inequality holds by the induction hypothesis.

(c) The second equality holds by the definition of the rollout algorithm.

(d) The third equality is the DP equation for the policy that corresponds to the base heuristic (this is the step where we need sequential consistency).

This completes the induction proof of the cost improvement property (6.16).
Q.E.D.

The cost improvement property of Prop. 6.4.1 may also be inferred by first transforming the finite horizon problem to an infinite horizon SSP problem, and then by using the cost improvement property of policy iteration for infinite horizon problems

Sequential Improvement

We will next show that the rollout policy has no worse performance than its base heuristic under a condition that is weaker than sequential consistency.

Let us recall that the rollout algorithm $\tilde{\pi} = \{\tilde{\mu}_0, \ldots, \tilde{\mu}_{N-1}\}$ is defined by the minimization

$$\tilde{\mu}_k(x_k) \in \arg \min_{u_k \in U_k(x_k)} \tilde{Q}_k(x_k, u_k),$$

where $\tilde{Q}_k(x_k, u_k)$ is the approximate Q-factor defined by

$$\tilde{Q}_k(x_k, u_k) = g_k(x_k, u_k) + H_{k+1}\big(f_k(x_k, u_k)\big),$$

[cf. Eq. (6.14)], and $H_{k+1}\big(f_k(x_k, u_k)\big)$ denotes the cost of the trajectory of the base heuristic starting from state $f_k(x_k, u_k)$.

Definition 6.4.2: We say that the base heuristic is *sequentially improving* if for all x_k and k, we have

$$\min_{u_k \in U_k(x_k)} \tilde{Q}_k(x_k, u_k) \leq H_k(x_k). \tag{6.18}$$

In words, the sequential improvement property (6.18) states that

Minimal heuristic Q-factor at x_k $\leq$ Heuristic cost at x_k.

Note that *when the heuristic is sequentially consistent it is also sequentially improving*. This follows from the preceding relation, since for a sequentially consistent heuristic, the heuristic cost at x_k is equal to the Q-factor of the control $\overline{u}_k$ that the heuristic applies at x_k,

$$\tilde{Q}_k(x_k, \overline{u}_k) = g_k(x_k, \overline{u}_k) + H_{k+1}\big(f_k(x_k, \overline{u}_k)\big),$$

which is greater or equal to the minimal Q-factor at x_k. This implies Eq. (6.18). We will now show that a sequentially improving heuristic yields policy improvement.

Proposition 6.4.2: (Cost Improvement Under Sequential Improvement) Consider the rollout policy $\tilde{\pi} = \{\tilde{\mu}_0, \ldots, \tilde{\mu}_{N-1}\}$ obtained with a sequentially improving base heuristic, and let $J_{k,\tilde{\pi}}(x_k)$ denote the cost obtained with $\tilde{\pi}$ starting from x_k at time k. Then

$$J_{k,\tilde{\pi}}(x_k) \leq H_k(x_k), \qquad \text{for all } x_k \text{ and } k,$$

where $H_k(x_k)$ denotes the cost of the base heuristic starting from x_k.

Proof: Follows from the calculation of Eq. (6.17), by replacing the last two steps (which rely on sequential consistency) with Eq. (6.18). **Q.E.D.**

Thus the rollout algorithm obtained with a sequentially improving base heuristic, will improve or at least will perform no worse than the base heuristic, from every starting state x_k. In fact *the algorithm has a monotonic improvement property, whereby it discovers a sequence of improved trajectories*. In particular, let us denote the trajectory generated by the base heuristic starting from x_0 by
$$T_0 = (x_0, u_0, \ldots, x_{N-1}, u_{N-1}, x_N),$$
and the final trajectory generated by the rollout algorithm starting from x_0 by
$$T_N = (x_0, \tilde{u}_0, \tilde{x}_1, \tilde{u}_1, \ldots, \tilde{x}_{N-1}, \tilde{u}_{N-1}, \tilde{x}_N).$$
Consider also the intermediate trajectories generated by the rollout algorithm given by
$$T_k = (x_0, \tilde{u}_0, \tilde{x}_1, \tilde{u}_1, \ldots, \tilde{x}_k, u_k, \ldots, x_{N-1}, u_{N-1}, x_N), \quad k = 1, \ldots, N-1,$$
where
$$(\tilde{x}_k, u_k, \ldots, x_{N-1}, u_{N-1}, x_N),$$
is the trajectory generated by the base heuristic starting from $\tilde{x}_k$. Then, by using the sequential improvement condition, it can be proved (see Fig. 6.4.4) that
$$\text{Cost of } T_0 \geq \cdots \geq \text{Cost of } T_k \geq \text{Cost of } T_{k+1} \geq \cdots \geq \text{Cost of } T_N. \quad (6.19)$$

Empirically, it has been observed that the cost improvement obtained by rollout with a sequentially improving heuristic is typically considerable and often dramatic. In particular, many case studies, dating to the middle 1990s, indicate consistently good performance of rollout; see the book [Ber20a] for a bibliography. The DP textbook [Ber17a] provides some detailed worked-out examples (Chapter 6, Examples 6.4.2, 6.4.5, 6.4.6, and Exercises 6.11, 6.14, 6.15, 6.16). The price for the performance improvement is extra computation that is typically equal to the computation time of the base heuristic times a factor that is a low order polynomial of N. It is generally hard to quantify the amount of performance improvement, but the computational results obtained from the case studies are consistent with the Newton step interpretations that we discussed in Chapter 3.

The books [Ber19a] (Section 2.5.1) and [Ber20a] (Section 3.1) show that the sequential improvement condition is satisfied in the context of MPC, and is the underlying reason for the stability properties of the MPC scheme. On the other hand the base heuristic underlying the classical form of the MPC scheme (cf. Section 5.3) is not sequentially consistent.

Generally, the sequential improvement condition may not hold for a given base heuristic. This is not surprising since any heuristic (no matter how inconsistent or silly) is in principle admissible to use as base heuristic. Here is an example:

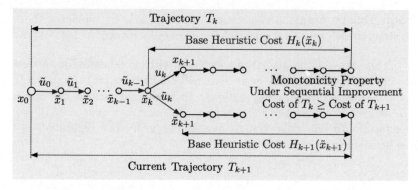

Figure 6.4.4 Proof of the monotonicity property (6.19). At $\tilde{x}_k$, the kth state generated by the rollout algorithm, we compare the "current" trajectory T_k whose cost is the sum of the cost of the current partial trajectory $(x_0, \tilde{u}_0, \tilde{x}_1, \tilde{u}_1, \ldots, \tilde{x}_k)$ and the cost $H_k(\tilde{x}_k)$ of the base heuristic starting from $\tilde{x}_k$, and the trajectory T_{k+1} whose cost is the sum of the cost of the partial rollout trajectory $(x_0, \tilde{u}_0, \tilde{x}_1, \tilde{u}_1, \ldots, \tilde{x}_k)$, and the Q-factor $\tilde{Q}_k(\tilde{x}_k, \tilde{u}_k)$ of the base heuristic starting from $(\tilde{x}_k, \tilde{u}_k)$. The sequential improvement condition guarantees that

$$H_k(\tilde{x}_k) \geq \tilde{Q}_k(\tilde{x}_k, \tilde{u}_k),$$

which implies that
$$\text{Cost of } T_k \geq \text{Cost of } T_{k+1}.$$

If strict inequality holds, the rollout algorithm will switch from T_k and follow T_{k+1}; cf. the traveling salesman example of Fig. 6.4.3.

Example 6.4.2 (Sequential Improvement Violation)

Consider the 2-stage problem shown in Fig. 6.4.5, which involves two states at each of stages 1 and 2, and the controls shown. Suppose that the unique optimal trajectory is $(x_0, u_0^*, x_1^*, u_1^*, x_2^*)$, and that the base heuristic produces this optimal trajectory starting at x_0. The rollout algorithm chooses a control at x_0 as follows: it runs the base heuristic to construct a trajectory starting from x_1^* and $\tilde{x}_1$, with corresponding costs $H_1(x_1^*)$ and $H_1(\tilde{x}_1)$. If

$$g_0(x_0, u_0^*) + H_1(x_1^*) > g_0(x_0, \tilde{u}_0) + H_1(\tilde{x}_1), \tag{6.20}$$

the rollout algorithm rejects the optimal control u_0^* in favor of the alternative control $\tilde{u}_0$. The inequality above will occur if the base heuristic chooses $\bar{u}_1$ at x_1^* (there is nothing to prevent this from happening, since the base heuristic is arbitrary), and moreover the cost $g_1(x_1^*, \bar{u}_1) + g_2(\tilde{x}_2)$, which is equal to $H_1(x_1^*)$ is high enough.

Let us also verify that if the inequality (6.20) holds then the heuristic is not sequentially improving at x_0, i.e., that

$$H_0(x_0) < \min\big\{g_0(x_0, u_0^*) + H_1(x_1^*),\ g_0(x_0, \tilde{u}_0) + H_1(\tilde{x}_1)\big\}.$$

Indeed, this is true because $H_0(x_0)$ is the optimal cost

$$H_0(x_0) = g_0(x_0, u_0^*) + g_1(x_1^*, u_1^*) + g_2(x_2^*),$$

Sec. 6.4 Rollout Algorithms for Discrete Optimization

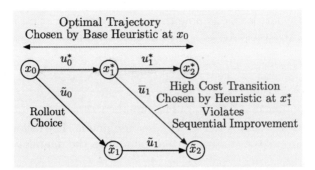

Figure 6.4.5 A 2-stage problem with states $x_1^*, \tilde{x}_1$ at stage 1, and states $x_2^*, \tilde{x}_2$ at stage 2. The controls and corresponding transitions are as shown in the figure. The rollout choice at the initial state x_0 is strictly suboptimal, while the base heuristic choice is optimal. The reason is that the base heuristic is not sequentially improving and makes the suboptimal choice $\overline{u}_1$ at x_1^*, but makes the different (optimal) choice u_1^* when run from x_0.

and must be smaller than both

$$g_0(x_0, u_0^*) + H_1(x_1^*),$$

which is the cost of the trajectory $(x_0, u_0^*, x_1^*, \overline{u}_1, \tilde{x}_2)$, and

$$g_0(x_0, \tilde{u}_0) + H_1(\tilde{x}_1),$$

which is the cost of the trajectory $(x_0, \tilde{u}_0, \tilde{x}_1, \tilde{u}_1, \tilde{x}_2)$.

The preceding example and the monotonicity property (6.19) suggest a simple enhancement to the rollout algorithm, which detects when the sequential improvement condition is violated and takes corrective measures. In this algorithmic variant, called *fortified rollout*, we maintain the best trajectory obtained so far, and keep following that trajectory up to the point where we discover another trajectory that has improved cost.

Using Multiple Base Heuristics - Parallel Rollout

In many problems, several promising heuristics may be available. It is then possible to use all of these heuristics in the rollout framework. The idea is to construct a *superheuristic*, which selects the best out of the trajectories produced by the entire collection of heuristics. The superheuristic can then be used as the base heuristic for a rollout algorithm.†

---

† A related practically interesting possibility is to introduce a partition of the state space into subsets, and a collection of multiple heuristics that are specially tailored to the subsets. We may then select the appropriate heuristic to use on each subset of the partition. In fact one may use a collection of multiple heuristics tailored to each subset of the state space partition, and at each state, select out of all the heuristics that apply, the one that yields minimum cost.

In particular, let us assume that we have m heuristics, and that the ℓth of these, given a state x_{k+1}, produces a trajectory

$$\tilde{T}^\ell_{k+1} = \{x_{k+1}, \tilde{u}^\ell_{k+1}, x_{k+2}, \ldots, \tilde{u}^\ell_{N-1}, \tilde{x}^\ell_N\},$$

and corresponding cost $C(\tilde{T}^\ell_{k+1})$. The superheuristic then produces at x_{k+1} the trajectory $\tilde{T}^\ell_{k+1}$ for which $C(\tilde{T}^\ell_{k+1})$ is minimum. The rollout algorithm selects at state x_k the control u_k that minimizes the minimal Q-factor:

$$\tilde{u}_k \in \arg\min_{u_k \in U_k(x_k)} \min_{\ell=1,\ldots,m} \tilde{Q}^\ell_k(x_k, u_k),$$

where

$$\tilde{Q}^\ell_k(x_k, u_k) = g_k(x_k, u_k) + C(\tilde{T}^\ell_{k+1})$$

is the cost of the trajectory $(x_k, u_k, \tilde{T}^\ell_{k+1})$. A similar idea was discussed in connection with MPC in Section 5.4, and in the paper [LJM21]. Note that the Q-factors of the different heuristics can be computed independently and in parallel. In view of this fact, the rollout scheme just described is sometimes referred to as *parallel rollout*.

An interesting property, which can be readily verified by using the definitions, is that *if all the heuristics are sequentially improving, the same is true for the superheuristic*, something that is also suggested by Fig. 6.4.4. Indeed, let us write the sequential improvement condition (6.18) for each of the base heuristics

$$\min_{u_k \in U_k(x_k)} \tilde{Q}^\ell_k(x_k, u_k) \leq H^\ell_k(x_k), \qquad \ell = 1, \ldots, m,$$

where $\tilde{Q}^\ell_k(x_k, u_k)$ and $H^\ell_k(x_k)$ are Q-factors and heuristic costs that correspond to the ℓth heuristic. Then by taking minimum over ℓ, we have

$$\min_{\ell=1,\ldots,m} \min_{u_k \in U_k(x_k)} \tilde{Q}^\ell_k(x_k, u_k) \leq \min_{\ell=1,\ldots,m} H^\ell_k(x_k),$$

for all x_k and k. By interchanging the order of the minimizations of the left side, we then obtain

$$\min_{u_k \in U_k(x_k)} \underbrace{\min_{\ell=1,\ldots,m} \tilde{Q}^\ell_k(x_k, u_k)}_{\text{Superheuristic Q-factor}} \leq \underbrace{\min_{\ell=1,\ldots,m} H^\ell_k(x_k)}_{\text{Superheuristic cost}},$$

which is precisely the sequential improvement condition (6.18) for the superheuristic.

Simplified Rollout Algorithms

We will now consider a rollout variant, called *simplified rollout algorithm*, which is motivated by problems where the control constraint set $U_k(x_k)$ is either infinite or finite but very large. Then the minimization

$$\tilde{\mu}_k(x_k) \in \arg\min_{u_k \in U_k(x_k)} \tilde{Q}_k(x_k, u_k), \tag{6.21}$$

[cf. Eqs. (6.13) and (6.14)], may be unwieldy, since the number of Q-factors

$$\tilde{Q}_k(x_k, u_k) = g_k(x_k, u_k) + H_{k+1}\big(f_k(x_k, u_k)\big)$$

is accordingly infinite or large.

To remedy this situation, we may replace $U_k(x_k)$ with a smaller finite subset $\overline{U}_k(x_k)$:

$$\overline{U}_k(x_k) \subset U_k(x_k).$$

The rollout control $\tilde{\mu}_k(x_k)$ in this variant is one that attains the minimum of $\tilde{Q}_k(x_k, u_k)$ over $u_k \in \overline{U}_k(x_k)$:

$$\tilde{\mu}_k(x_k) \in \arg\min_{u_k \in \overline{U}_k(x_k)} \tilde{Q}_k(x_k, u_k). \tag{6.22}$$

An example is when $\overline{U}_k(x_k)$ results from discretization of an infinite set $U_k(x_k)$. Another possibility is when by using some preliminary approximate optimization, we can identify a subset $\overline{U}_k(x_k)$ of promising controls, and to save computation, we restrict attention to this subset. A related possibility is to generate $\overline{U}_k(x_k)$ by some random search method that explores intelligently the set $U_k(x_k)$ with the aim to minimize $\tilde{Q}_k(x_k, u_k)$ [cf. Eq. (6.21)].

It turns out that the proof of the cost improvement property of Prop. 6.4.2,

$$J_{k,\tilde{\pi}}(x_k) \leq H_k(x_k), \qquad \text{for all } x_k \text{ and } k,$$

goes through if the following modified sequential improvement property holds:

$$\min_{u_k \in \overline{U}_k(x_k)} \tilde{Q}_k(x_k, u_k) \leq H_k(x_k). \tag{6.23}$$

This can be seen by verifying that Eq. (6.23) is sufficient to guarantee that the monotone improvement Eq. (6.19) is satisfied. The condition (6.23) is very simple to satisfy if the base heuristic is sequentially consistent, in which case the control $\overline{u}_k$ selected by the base heuristic satisfies

$$\tilde{Q}_k(x_k, \overline{u}_k) = H_k(x_k).$$

In particular, for the property (6.23) to hold, it is sufficient that $\overline{U}_k(x_k)$ contains the base heuristic choice $\overline{u}_k$.

The idea of replacing the minimization (6.21) by the simpler minimization (6.22) can be extended. In particular, by working through the preceding argument, it can be seen that *any policy*

$$\tilde{\pi} = \{\tilde{\mu}_0, \ldots, \tilde{\mu}_{N-1}\}$$

such that $\tilde{\mu}_k(x_k)$ satisfies the condition

$$\tilde{Q}_k(x_k, \tilde{\mu}_k(x_k)) \leq H_k(x_k),$$

for all x_k and k, guarantees the modified sequential improvement property (6.23), and hence also the cost improvement property. A prominent example of such an algorithm arises in the multiagent case where u has m components, $u = (u^1, \ldots, u^m)$, and the minimization over $U_k^1(x_k) \times \cdots \times U_k^m(x_k)$ is replaced by a sequence of single component minimizations, one-component-at-a-time; cf. Section 3.7.

The Fortified Rollout Algorithm

In this section we describe a rollout variant that implicitly enforces the sequential improvement property. This variant, called the *fortified rollout algorithm*, starts at x_0, and generates step-by-step a sequence of states $\{x_0, x_1, \ldots, x_N\}$ and corresponding sequence of controls. Upon reaching state x_k we have the trajectory

$$\overline{P}_k = \{x_0, u_0, \ldots, u_{k-1}, x_k\}$$

that has been constructed by rollout, called *permanent trajectory*, and we also store a *tentative best trajectory*

$$\overline{T}_k = \{x_0, u_0, \ldots, u_{k-1}, x_k, \overline{u}_k, \overline{x}_{k+1}, \overline{u}_{k+1}, \ldots, \overline{u}_{N-1}, \overline{x}_N\}$$

with corresponding cost

$$C(\overline{T}_k) = \sum_{t=0}^{k-1} g_t(x_t, u_t) + g_k(x_k, \overline{u}_k) + \sum_{t=k+1}^{N-1} g_t(\overline{x}_t, \overline{u}_t) + g_N(\overline{x}_N).$$

The tentative best trajectory $\overline{T}_k$ is the best end-to-end trajectory computed up to stage k of the algorithm. Initially, $\overline{T}_0$ is the trajectory generated by the base heuristic starting at the initial state x_0. The idea now is to *discard the suggestion of the rollout algorithm at every state x_k where it produces a trajectory that is inferior to $\overline{T}_k$, and use $\overline{T}_k$ instead* (see Fig. 6.4.6).†

---

† The fortified rollout algorithm can actually be viewed as the ordinary rollout algorithm applied to a modified version of the original problem and modified base heuristic that has the sequential improvement property. This construction is somewhat technical and unintuitive and will not be given; we refer to Bertsekas, Tsitsiklis, and Wu [BTW97], and the DP textbook [Ber17a], Section 6.4.2.

Sec. 6.4 Rollout Algorithms for Discrete Optimization

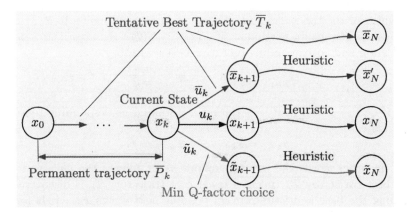

Figure 6.4.6 Schematic illustration of fortified rollout. After k steps, we have constructed the permanent trajectory

$$\overline{P}_k = \{x_0, u_0, \ldots, u_{k-1}, x_k\},$$

and the tentative best trajectory

$$\overline{T}_k = \{x_0, u_0, \ldots, u_{k-1}, x_k, \overline{u}_k, \overline{x}_{k+1}, \overline{u}_{k+1}, \ldots, \overline{u}_{N-1}, \overline{x}_N\},$$

the best end-to-end trajectory computed so far. We now run the rollout algorithm at x_k, i.e., we find the control $\tilde{u}_k$ that minimizes over u_k the sum of $g_k(x_k, u_k)$ plus the heuristic cost from the state $x_{k+1} = f_k(x_k, u_k)$, and the corresponding trajectory

$$\tilde{T}_k = \{x_0, u_0, \ldots, u_{k-1}, x_k, \tilde{u}_k, \tilde{x}_{k+1}, \tilde{u}_{k+1}, \ldots, \tilde{u}_{N-1}, \tilde{x}_N\}.$$

If the cost of the end-to-end trajectory $\tilde{T}_k$ is lower than the cost of $\overline{T}_k$, we add $(\tilde{u}_k, \tilde{x}_{k+1})$ to the permanent trajectory and set the tentative best trajectory to $\overline{T}_{k+1} = \tilde{T}_k$. Otherwise we add $(\overline{u}_k, \overline{x}_{k+1})$ to the permanent trajectory and keep the tentative best trajectory unchanged: $\overline{T}_{k+1} = \overline{T}_k$.

In particular, upon reaching state x_k, we run the rollout algorithm as earlier, i.e., for every $u_k \in U_k(x_k)$ and next state $x_{k+1} = f_k(x_k, u_k)$, we run the base heuristic from x_{k+1}, and find the control $\tilde{u}_k$ that gives the best trajectory, denoted

$$\tilde{T}_k = \{x_0, u_0, \ldots, u_{k-1}, x_k, \tilde{u}_k, \tilde{x}_{k+1}, \tilde{u}_{k+1}, \ldots, \tilde{u}_{N-1}, \tilde{x}_N\}$$

with corresponding cost

$$C(\tilde{T}_k) = \sum_{t=0}^{k-1} g_t(x_t, u_t) + g_k(x_k, \tilde{u}_k) + \sum_{t=k+1}^{N-1} g_t(\tilde{x}_t, \tilde{u}_t) + g_N(\tilde{x}_N).$$

Whereas the ordinary rollout algorithm would choose control $\tilde{u}_k$ and move to $\tilde{x}_{k+1}$, the fortified algorithm compares $C(\overline{T}_k)$ and $C(\tilde{T}_k)$, and depending on which of the two is smaller, chooses $\overline{u}_k$ or $\tilde{u}_k$ and moves to $\overline{x}_{k+1}$ or to $\tilde{x}_{k+1}$, respectively. In particular, if $C(\overline{T}_k) \leq C(\tilde{T}_k)$ the algorithm sets the next state and corresponding tentative best trajectory to

$$x_{k+1} = \overline{x}_{k+1}, \qquad \overline{T}_{k+1} = \overline{T}_k,$$

and if $C(\overline{T}_k) > C(\tilde{T}_k)$ it sets the next state and corresponding tentative best trajectory to

$$x_{k+1} = \tilde{x}_{k+1}, \qquad \overline{T}_{k+1} = \tilde{T}_k.$$

In other words the fortified rollout at x_k follows the current tentative best trajectory $\overline{T}_k$ unless a lower cost trajectory $\tilde{T}_k$ is discovered by running the base heuristic from all possible next states x_{k+1}.† It follows that at every state the tentative best trajectory has no larger cost than the initial tentative best trajectory, which is the one produced by the base heuristic starting from x_0. Moreover, it can be seen that if the base heuristic is sequentially improving, the rollout algorithm and its fortified version coincide. Experimental evidence suggests that it is often important to use the fortified version if the base heuristic is not known to be sequentially improving. Fortunately, the fortified version involves hardly any additional computational cost.

As expected, when the base heuristic generates an optimal trajectory, the fortified rollout algorithm will also generate the same trajectory. This is illustrated by the following example.

Example 6.4.3

Let us consider the application of the fortified rollout algorithm to the problem of Example 6.4.2 and see how it addresses the issue of cost improvement. The fortified rollout algorithm stores as initial tentative best trajectory the optimal trajectory $(x_0, u_0^*, x_1^*, u_1^*, x_2^*)$ generated by the base heuristic at x_0. Then, starting at x_0, it runs the heuristic from x_1^* and $\tilde{x}_1$, and (despite the fact that the ordinary rollout algorithm prefers going to $\tilde{x}_1$ rather than x_1^*) it discards the control $\tilde{u}_0$ in favor of u_0^*, which is dictated by the tentative best trajectory. It then sets the tentative best trajectory to $(x_0, u_0^*, x_1^*, u_1^*, x_2^*)$.

We finally note that the fortified rollout algorithm can be used in a different setting to restore and maintain the cost improvement property. Suppose in particular that the rollout minimization at each step is performed with approximations. For example the control u_k may have multiple independently constrained components, i.e.,

$$u_k = (u_k^1, \ldots, u_k^m), \qquad U_k(x_k) = U_k^1(x_k) \times \cdots \times U_k^m(x_k).$$

† The base heuristic may also be run from a subset of the possible next states x_{k+1}, as in the case where a simplified version of rollout is used. Then fortified rollout will still guarantee a cost improvement property.

Sec. 6.4　　Rollout Algorithms for Discrete Optimization

Then, to take advantage of distributed computation, it may be attractive to decompose the optimization over u_k in the rollout algorithm,

$$\tilde{\mu}_k(x_k) \in \arg\min_{u_k \in U_k(x_k)} \Big[g_k(x_k, u_k) + H_{k+1}\big(f_k(x_k, u_k)\big)\Big],$$

into an (approximate) parallel optimization over the components u_k^i (or subgroups of these components). However, as a result of approximate optimization over u_k, the cost improvement property may be degraded, even if the sequential improvement assumption holds. In this case by maintaining the tentative best trajectory, starting with the one produced by the base heuristic at the initial condition, we can ensure that the fortified rollout algorithm, even with approximate minimization, will not produce an inferior solution to the one of the base heuristic.

Model-Free Rollout

We will now consider a rollout algorithm for discrete deterministic optimization for the case where *we do not know the cost function and the constraints of the problem*. Instead we have access to a base heuristic, and also a human or software "expert" who can rank any two feasible solutions without assigning numerical values to them.

We consider the general discrete optimization problem of selecting a control sequence $u = (u_0, \ldots, u_{N-1})$ to minimize a function $G(u)$. For simplicity we assume that each component u_k is constrained to lie in a given constraint set U_k, but extensions to more general constraint sets are possible. We assume the following:

(a) A base heuristic with the following property is available: Given any $k < N-1$, and a partial solution $(u_0, \ldots, u_k)$, it generates, for every $\tilde{u}_{k+1} \in U_{k+1}$, a complete feasible solution by concatenating the given partial solution $(u_0, \ldots, u_k)$ with a sequence $(\tilde{u}_{k+1}, \ldots, \tilde{u}_{N-1})$. This complete feasible solution is denoted

$$S_k(u_0, \ldots, u_k, \tilde{u}_{k+1}) = (u_0, \ldots, u_k, \tilde{u}_{k+1}, \ldots, \tilde{u}_{N-1}).$$

The base heuristic is also used to start the algorithm from an artificial empty solution, by generating all components $\tilde{u}_0 \in U_0$ and a complete feasible solution $(\tilde{u}_0, \ldots, \tilde{u}_{N-1})$, starting from each $\tilde{u}_0 \in U_0$.

(b) An "expert" is available that can compare any two feasible solutions u and $\overline{u}$, in the sense that he/she can determine whether

$$G(u) > G(\overline{u}), \quad \text{or} \quad G(u) \leq G(\overline{u}).$$

It can be seen that deterministic rollout can be applied to this problem, even though the cost function G is unknown. The reason is that the

148 Finite Horizon Deterministic Problems - Discrete Optimization   Chap. 6

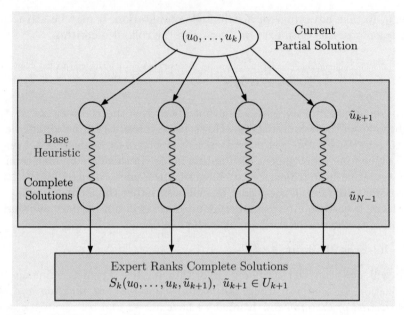

Figure 6.4.7 Schematic illustration of model-free rollout with an expert for minimizing $G(u)$ subject to $u \in U_0 \times \cdots \times U_{N-1}$. We assume that we do not know G and/or $U_0, \ldots, U_{N-1}$. Instead we have a base heuristic, which given a partial solution $(u_0, \ldots, u_k)$, outputs all next controls $\tilde{u}_{k+1} \in U_{k+1}$, and generates from each a complete solution

$$S_k(u_0, \ldots, u_k, \tilde{u}_{k+1}) = (u_0, \ldots, u_k, \tilde{u}_{k+1}, \ldots, \tilde{u}_{N-1}).$$

Also, we have a human or software "expert" that can rank any two complete solutions without assigning numerical values to them. The control that is selected from U_{k+1} by the rollout algorithm is the one whose corresponding complete solution is ranked best by the expert.

rollout algorithm uses the cost function only as a means of ranking complete solutions in terms of their cost. Hence, if the ranking of any two solutions can be revealed by the expert, this is all that is needed.† In fact, the constraint sets $U_0, \ldots, U_{N-1}$ need not be known either, as long as they can be generated by the base heuristic. Thus, the rollout algorithm can be described as follows (see Fig. 6.4.7):

We start with an artificial empty solution, and at the typical step, given the partial solution $(u_0, \ldots, u_k)$, $k < N-1$, we use the base heuristic

---

† Note that for this to be true, it is important that the problem is deterministic, and that the expert ranks solutions using some underlying (though unknown) cost function. In particular, the expert's rankings should have a transitivity property: if u is ranked better than u' and u' is ranked better than u'', then u is ranked better than u''.

Sec. 6.5   Approximation in Value Space with Multistep Lookahead   149

to generate all possible one-step-extended solutions

$$(u_0, \ldots, u_k, \tilde{u}_{k+1}), \qquad \tilde{u}_{k+1} \in U_{k+1},$$

and the set of complete solutions

$$S_k(u_0, \ldots, u_k, \tilde{u}_{k+1}), \qquad \tilde{u}_{k+1} \in U_{k+1}.$$

We then use the expert to rank this set of complete solutions. Finally, we select the component u_{k+1} that is ranked best by the expert, extend the partial solution $(u_0, \ldots, u_k)$ by adding u_{k+1}, and repeat with the new partial solution $(u_1, \ldots, u_k, u_{k+1})$.

Except for the (mathematically inconsequential) use of an expert rather than a cost function, the preceding rollout algorithm can be viewed as a special case of the one given earlier. As a result several of the rollout variants that we have discussed so far (rollout with multiple heuristics, simplified rollout, and fortified rollout) can also be easily adapted. For an application of the model-free rollout methodology to the problem of RNA folding, see the paper [LPS21] and the RL book [Ber20a].

6.5 ROLLOUT AND APPROXIMATION IN VALUE SPACE WITH MULTISTEP LOOKAHEAD

We will now consider incorporating multistep lookahead minimization into the rollout framework. To describe two-step lookahead for deterministic problems, suppose that after k steps we have reached state x_k. We then consider the set of all possible two-step-ahead states x_{k+2}, we run the base heuristic starting from each of them, and compute the two-stage cost to get from x_k to x_{k+2}, plus the cost of the base heuristic from x_{k+2}. We select the state, say $\tilde{x}_{k+2}$, that is associated with minimum cost, compute the controls $\tilde{u}_k$ and $\tilde{u}_{k+1}$ that lead from x_k to $\tilde{x}_{k+2}$, choose $\tilde{u}_k$ as the next rollout control and $x_{k+1} = f_k(x_k, \tilde{u}_k)$ as the next state, and discard $\tilde{u}_{k+1}$.

The extension of the algorithm to lookahead of more than two steps is straightforward: instead of the two-step-ahead states x_{k+2}, we run the base heuristic starting from all the possible ℓ-step ahead states $x_{k+\ell}$, etc; see Fig. 6.5.1. For cases where the ℓ-step lookahead minimization is very time consuming, we may consider variants involving approximations aimed at simplifying the associated computations.

An important variation for problems with a long horizon is *truncated rollout with terminal cost approximation*. Here the rollout trajectories are obtained by running the base heuristic from the leaf nodes of the lookahead graph, and they are truncated after a given number of steps, while a terminal cost approximation is added to the heuristic cost to compensate for the resulting error; see Fig. 6.5.2. One possibility that works well for many problems, particularly when the combined lookahead for minimization and

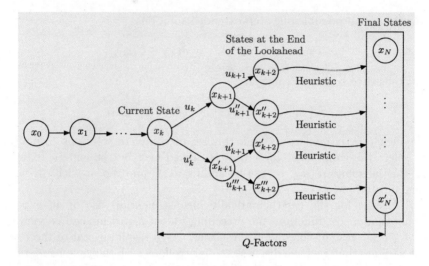

Figure 6.5.1 Illustration of multistep rollout with $\ell = 2$ for deterministic problems. We run the base heuristic from each leaf $x_{k+\ell}$ at the end of the lookahead graph. We then construct an optimal solution for the lookahead minimization problem, where the heuristic cost is used as terminal cost approximation. We thus obtain an optimal ℓ-step control sequence through the lookahead graph, use the first control in the sequence as the rollout control, discard the remaining controls, move to the next state, and repeat. Note that the multistep lookahead minimization may involve approximations aimed at simplifying the associated computations.

base heuristic simulation is long, is to simply set the terminal cost approximation to zero. Alternatively, the terminal cost function approximation can be obtained by problem approximation or by using some sophisticated off-line training process that may involve an approximation architecture such as a neural network.

An important observation is that the preceding algorithmic scheme can be viewed as multistep approximation in value space, and *it can be interpreted as a Newton step*, with suitable starting point that is determined by the truncated rollout with the base heuristic, and the terminal cost approximation. This interpretation is possible once the discrete optimal control problem is reformulated to an equivalent infinite horizon SSP problem; cf. the discussion of Section 6.1. Thus the algorithm inherits the fast convergence property of the Newton step, which we have discussed in the context of infinite horizon problems.

6.5.1 Simplified Multistep Rollout - Double Rollout

The main difficulty in applying rollout with ℓ-step lookahead stems from the rapid expansion of the lookahead graph as ℓ increases, and the accordingly large number of applications of the base heuristic. This difficulty also arises

Sec. 6.5 Approximation in Value Space with Multistep Lookahead

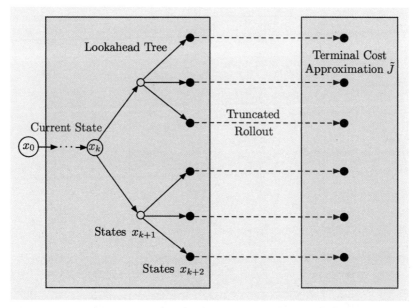

Figure 6.5.2 Illustration of truncated rollout with two-step lookahead and a terminal cost approximation $\tilde{J}$. The base heuristic is used for a limited number of steps and the terminal cost is added to compensate for the remaining steps.

more generally in approximation in value space schemes with multistep lookahead. In these circumstances, we may consider *simplified rollout*, which is based on "pruning" the lookahead graph, so that the number of its nodes becomes manageable. By this we mean disregarding some of the states that are ℓ steps or less ahead, which are judged less promising according to some criterion (for example the costs of the base heuristic after a one-step or a short lookahead); see Fig. 6.5.3. This is also reminiscent of the Monte Carlo Tree Search (MCTS) technique, which has been used in AlphaZero, but we have not discussed in this book; descriptions can be found in several of the textbooks that we already have referenced.

Simplified rollout aims to limit the number of times that the base heuristic is applied, which can grow exponentially as the length of lookahead is increased. In some contexts, it may also be viewed as *selective depth lookahead*, whereby the lookahead graph is expanded nonuniformly (lookahead is deeper from some states than for others).

An interesting idea for selective depth lookahead and tree pruning, which can also be combined with simplified rollout, is based on applying rollout to the solution of the ℓ-step lookahead minimization; after all, this is also a discrete optimization problem that can be addressed by any suboptimal method, including rollout. Thus we may use a *second base heuristic* to generate a promising trajectory through the ℓ-step lookahead tree by using m-step lookahead rollout where $1 < m < \ell$. This "double rollout"

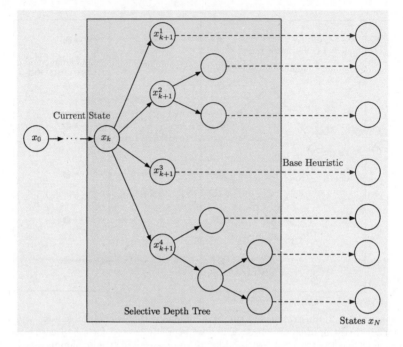

Figure 6.5.3 Illustration of a form of deterministic rollout with selective depth lookahead. After k steps of the algorithm, we have a trajectory that starts at the initial state x_0 and ends at state x_k. We then generate the set of all possible next states (states x_{k+1}^1, x_{k+1}^2, x_{k+1}^3, x_{k+1}^4 in the figure). We "evaluate" these states using the base heuristic, and select some of them for "expansion," i.e., we generate their next states x_{k+2}, evaluate them using the base heuristic, and continue. In the end we have a selective depth graph of next states, and the base heuristic costs starting from the nodes of the graph. The state x_{k+1} that corresponds to the smallest overall cost is chosen by the selective depth lookahead rollout algorithm. For problems with a large number of stages, we can also truncate the rollout trajectories and add a terminal cost function approximation as compensation for the resulting error; cf. Fig. 6.5.2.

algorithm requires a number of heuristic applications that grows *linearly rather than exponentially* with ℓ. In particular, at each stage, *the number of heuristic applications of the ℓ-step rollout and of the "double rollout" algorithm will be bounded by n^ℓ and by $n^{m+1} \cdot (\ell - m)$*, respectively, where n is a bound on the number of control choices at each state. Note that the two base heuristics need not be related to each other, as they are applied to different problems. Moreover, the idea of solving the ℓ-step lookahead minimization problem approximately by a form of truncated rollout applies more generally, to approximation in value space with multistep lookahead.

The double rollout method admits several variations. These include a *fortified version*, which guards against lack of sequential improvement, and the possibility that the rollout algorithm gets sidetracked along an inferior

Sec. 6.5   Approximation in Value Space with Multistep Lookahead   153

trajectory. The fortified algorithm maintains a tentative best trajectory, from which it will not deviate until the rollout algorithm generates a less costly trajectory, similar to the one-step lookahead case.

Let us also mention a variant that maintains multiple trajectories, extending from a given state x_k to possibly multiple next states x_{k+1}, and subsequent states $x_{k+2}, x_{k+3}, \ldots$. These states are the ones considered "most promising" based on the current results of the multistep minimization (like being "ϵ-best"), but may be discarded later based on subsequent computations. Such extended forms of rollout can be combined with a fortified rollout scheme to ensure cost improvement over the base heuristic. They are restricted to deterministic problems, and tend to be problem-dependent. The idea of exploring multiple trajectories can be implemented systematically through an incremental variant of double rollout, which we discuss next.

6.5.2   Incremental Rollout for Multistep Approximation in Value Space

We will now consider an extension of the double rollout scheme, which we call *incremental rollout*. It can be used to solve approximately the lookahead minimization problem that arises in multistep approximation in value space with arbitrary terminal cost function approximation (not necessarily obtained by some form of truncated rollout). At a given state, it applies shortest path-like computations to a sequence of subgraphs of a multistep lookahead graph, with the size of the subgraphs expanding iteratively. The difference from double rollout is the following: in incremental rollout a subgraph of multiple paths is iteratively extended starting from the current state going towards the end of the lookahead horizon, instead of extending a single path as in double rollout. This is similar to what is done in Monte Carlo Tree Search (MCTS), which is also designed to solve approximately general multistep lookahead minimization problems (including stochastic ones), and involves iterative expansion of an acyclic lookahead graph to new nodes, as well as backtracking to previously encountered nodes.

Aside from expedited computation, another advantage of incremental rollout is that it has the character of an "anytime" algorithm. By this we mean "an algorithm that can return a feasible solution even if it is interrupted before it ends" (as defined by Wikipedia). This is a shared attraction with MCTS. However, incremental rollout seems to be more appropriate than MCTS for deterministic problems, where there are no random variables in the problem's model and therefore Monte Carlo simulation does not make sense.

More specifically, the objective of incremental rollout is to find a good approximate solution of the ℓ-step lookahead minimization problem that arises in approximation in value space schemes for finite or infinite horizon deterministic DP problems. To describe this problem, we have an acyclic

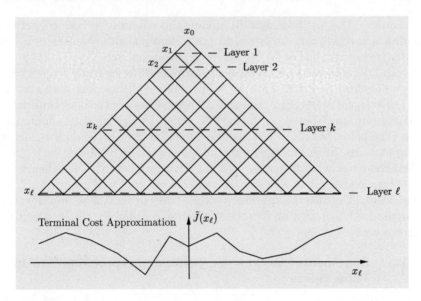

Figure 6.5.4 Illustration of the ℓ-step lookahead minimization problem. We have an acyclic graph with a single root (the current state x_0) and ℓ layers, with the kth layer consisting of the states x_k that are reachable from x_0 with a feasible sequence of k controls. For the states x_ℓ of the last layer there is a given terminal cost approximation $\tilde{J}(x_\ell)$.

graph with a single root (the current state, denoted x_0) and ℓ layers, with the kth layer consisting of the states x_k that are reachable from x_0 with a feasible sequence of k controls. There is an arc for every state x_1 of the 1st layer that can be reached from x_0 with a feasible control, and similarly an arc for every pair of states (x_k, x_{k+1}), layers k and $k+1$, respectively, for which x_{k+1} can be reached from x_k with a feasible control. The cost of each of these arcs is the stage cost of the corresponding state-control pair, minimized over all possible controls that correspond to the same pair (x_k, x_{k+1}). Mathematically, the cost of the arc (x_k, x_{k+1}) is

$$\hat{g}_k(x_k, x_{k+1}) = \min_{\{u_k \in U_k(x_k) \;|\; x_{k+1} = f_k(x_k, u_k)\}} g_k(x_k, u_k). \qquad (6.24)$$

For the states x_ℓ of the last layer there is also a given terminal cost approximation $\tilde{J}(x_\ell)$, possibly obtained through off-line training. It can be thought of as the cost of an artificial arc connecting x_ℓ to an artificial termination state; see Fig. 6.5.4.

The ℓ-step lookahead minimization problem just described is a shortest path problem that can be solved exactly by forward DP. This is a method that successively finds the shortest distances from x_0 to the nodes $x_1, \ldots, x_\ell$ of layers $1, \ldots, \ell$, taking also into account the terminal cost approximation $\tilde{J}(x_\ell)$. The method has an incremental character, with each

Sec. 6.5   Approximation in Value Space with Multistep Lookahead   155

iteration solving a shortest path problem that starts from x_0 and involves the states of one additional layer: first the shortest path problem from x_0 to the states x_1, then from x_0 to the states x_2, using the shortest distances from x_0 to x_1, etc. The results of each iteration are used to expedite the computations of the next iteration. In particular, the shortest distances from x_0 to the states x_{k+1} of layer $k+1$, denoted $D_{k+1}(x_{k+1})$, are obtained from

$$D_{k+1}(x_{k+1}) = \min_{x_k \in \text{ (layer } k)} \{D_k(x_k) + \hat{g}_k(x_k, x_{k+1})\}, \tag{6.25}$$

where $\hat{g}_k(x_k, x_{k+1})$ is given by Eq. (6.24), and $D_k(x_k)$ is the shortest distance from x_0 to x_k. This is the form of the classical forward DP algorithm for ℓ-step deterministic DP problems (see e.g., the author's DP textbook [Ber17a], Section 2.1).

Like the forward DP algorithm, the incremental rollout algorithm that we will consider next also expands the length of the shortest path computation iteratively. Every iteration uses a rollout-like computation to produce a control sequence of the form $\{u_0, \ldots, u_{k-1}\}$, from x_0 to some state x_k, where $1 \le k \le \ell$. At the last iteration, i.e., upon the algorithm's termination, the first element u_0 of the corresponding sequence is applied at the current state x_0. However, the method need not generate sequences with increasing number of elements, like the forward DP algorithm described above. Instead it can occasionally backtrack to generate a control sequence with fewer controls.

The algorithm can terminate at any time after the first iteration, as long as the number of components of the generated sequence does not exceed ℓ. Thus it may be used with very large ℓ, and keep searching to ever larger depth for as long as real-time computation constraints permit. If allowed to run long enough, the algorithm will eventually perform an ℓ-step lookahead minimization with a portion of the lookahead graph pruned out, i.e., with selective length lookahead. Generally, the algorithm tends to find better solutions as it progresses, particularly when combined with some form of fortified rollout. Moreover, the Newton step interpretation still applies to the algorithm.

Incremental Rollout Algorithm Description

We will now describe in more detail our incremental rollout algorithm. It aims to find an approximately optimal path $(x_0, x_1, \ldots, x_\ell)$ through the ℓ-step lookahead graph, which we refer to as G. At the start of each iteration, we have an acyclic subgraph S of G, which contains x_0. We also have a path P that starts at x_0, goes through S, and ends at one of the terminal nodes of S (the ones with no outgoing arcs). Figure 6.5.5 provides an illustration.

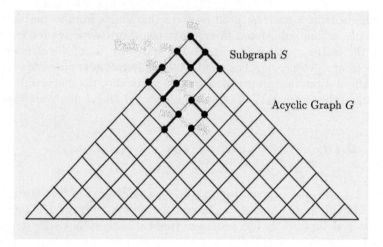

Figure 6.5.5 Illustration of the starting data of an iteration of the incremental rollout algorithm. The iteration starts with a connected acyclic subgraph S of the given multistep lookahead graph G, which contains x_0, and with a path P that starts at x_0 and goes through S and ends at one of the terminal nodes of S (the ones with no outgoing arcs). In this figure, P is the path $(x_0, x_1, x_2, x_3, x_4, x_5, x_6)$, while S consists of the nodes that are marked with circles.

The algorithm starts with a connected acyclic subgraph S of the given multistep lookahead graph G, which contains x_0, and with a path P that starts at x_0, goes through S and ends at one of the terminal nodes of S (one possibility is take S and P equal to just the root node x_0). It terminates when the end node of P belongs to layer ℓ (or earlier if a computational budget constraint is reached). Consistent with the rollout methodology, the algorithm uses a base heuristic, which is possibly obtained through off-line training. This heuristic has the property that for every k and node x_k of layer k, it can generate a path from x_k to a node x_ℓ of layer ℓ. The sum of transition costs along this path, plus the terminal cost $\tilde{J}(x_\ell)$, is denoted by $H(x_k)$. The algorithm also uses a scalar $\delta \geq 0$, which together with the base heuristic and the subgraph S, defines the path P based on a shortest path computation. In particular, P is defined as the path from x_0 to one of the terminal nodes x of S (the nodes of S with no successors in S) that minimizes the

Sum of stage costs along P from x_0 to x
$+ \delta \cdot$ (Number of arcs of P)
$+ H(x)$.

Thus P is a shortest path, assuming that each arc (x_k, x_{k+1}) along the path has length equal to δ plus the corresponding stage cost $\hat{g}(x_k, x_{k+1})$ given by Eq. (6.24), and a terminal cost equal to $H(x)$ is added. When $\delta = 0$, the cost of P is the sum of the arc lengths/stage costs $\hat{g}(x_k, x_{k+1})$,

Sec. 6.5   Approximation in Value Space with Multistep Lookahead

plus the base heuristic cost $H(x)$. When $\delta > 0$, an additional cost of size δ is added to the cost of each arc traversed, thus encouraging paths with fewer arcs.

At an iteration starting with S and $P = (x_0, x_1, \ldots, x_k)$, with x_k not in layer ℓ, we perform two operations that define the new subgraph $\overline{S}$ and new path $\overline{P}$:

(a) An *expansion of S*, whereby we add to S the outgoing arcs from x_k (the terminal node of P) to form a new acyclic subgraph $\overline{S}$.

(b) A *shortest path computation*, whereby we obtain $\overline{P}$ as a shortest path within the new subgraph $\overline{S}$, with arc lengths as described above.

Figure 6.5.6 provides an illustration. Note that the new path $\overline{P}$ may involve just an expansion of the current path P by a single arc, as in Fig. 6.5.6(b), but may also be very different from P, as in Fig. 6.5.6(c), where the iteration "backtracks" to some node added to the tree at an earlier iteration. A key fact is that the shortest path computation within $\overline{S}$ is very simple given the results of the preceding shortest path computation that obtained S, similar to the forward DP algorithm (6.25). This involves incremental changes of shortest distances of nodes of P, to reflect the effect of the nodes added to S though the expansion operation, as can be readily verified by the reader.

The role of the parameter δ is noteworthy. When $\delta = 0$ and the base heuristic is sequentially improving, it can be seen that the incremental rollout algorithm performs exactly like the double rollout algorithm, and extends a single path from x_0 to layer ℓ, adding a single node at each iteration. On the other hand when δ is large enough, the algorithm operates like the forward DP algorithm (6.25). The reason is that a very large value of δ forces the algorithm to visit all nodes of a given layer before proceeding to the next layer, since each additional arc in P incurs an extra cost of δ.

Generally, as δ increases, the algorithm tends to backtrack more often, and to generate more paths through the graph, thereby visiting more nodes and increasing the number of applications of the base heuristic. Thus δ may be viewed as an *exploration parameter*; when δ is large the algorithm tends to explore more paths thereby improving the quality of the multistep lookahead minimization, at the expense of greater computational effort. In the absence of additional problem-specific information, favorable values of δ should be obtained through experimentation. On the other hand, one may consider adaptive schemes, whereby δ may depend on the current state x_0. For example, one may increase or decrease δ depending on some measure of size of S. In particular, let K be the number of layers k such that S contains some nodes of layer k, and consider the following ratio

$$R_k = \frac{\text{number of nodes of layer } k \text{ that belong to } S}{\text{number of nodes in layer } k},$$

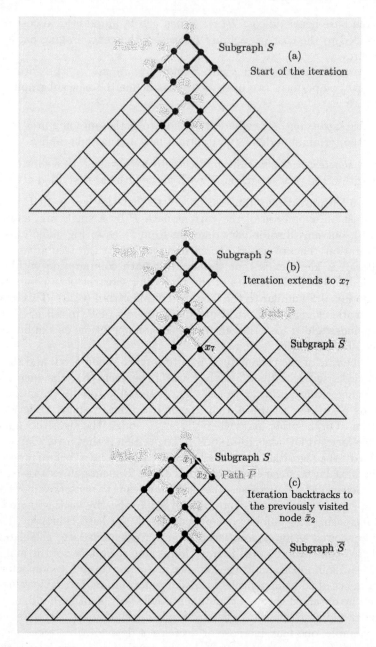

Figure 6.5.6 Illustration of an iteration of the incremental rollout algorithm. The iteration starts with S and P as in figure (a); cf. Fig. 6.5.5. During the iteration the outgoing arcs from the terminal node x_6 are added to S in order to form $\overline{S}$ as in figures (b) and (c). A shortest path $\overline{P}$ is then computed from x_0 to the terminal nodes of $\overline{S}$. There are two possibilities: (1) $\overline{P}$ consists of P plus an additional outgoing arc from x_6, as in figure (b) where the new arc added to P is (x_6, x_7), and the terminal node of $\overline{P}$ does not belong to S. (2) The last arc of $\overline{P}$ is not an outgoing arc of the terminal node of P, as in figure (c), where the iteration backtracks to node $\bar{x}_2$, which already belongs to S, and $\overline{P} = (x_0, \bar{x}_1, \bar{x}_2)$.

for each layer $k = 1, \ldots, K$, as well as the "average ratio"

$$R = \frac{1}{K} \sum_{k=1}^{K} R_k.$$

In an adaptive scheme for adjusting δ, we may keep track of the ratio R as the algorithm progresses, and adjust δ accordingly: increase δ if R is relatively small, and decrease δ if R is relatively large.

It is interesting to compare incremental rollout with the A^* algorithm, a well-known method for searching deterministic graphs, which also involves expansions to successor nodes, shortest path-like computations, and a "heuristic distance." In A^* when a node is expanded, eventually all its successors are checked for improvement of their distance from the root, in which case they are entered in a candidate list for future expansion. In incremental rollout, only the node of shortest total distance is expanded at each iteration. Thus A^* expands many more nodes than incremental rollout. As an indication of the difference between the two methods, we note that A^* will find the shortest path from the root to layer ℓ under some conditions on the base heuristic. Incremental rollout offers the same guarantee only for large enough values of δ. Moreover, incremental rollout is well suited to the layered structure of the multistep lookahead graph, whereas A^* can be used for any deterministic shortest path problem, whether it involves an acyclic graph or not.

6.6 CONSTRAINED FORMS OF ROLLOUT ALGORITHMS

In this section we will discuss constrained deterministic DP problems, including challenging combinatorial optimization and integer programming problems. We introduce a rollout algorithm, which relies on a base heuristic and applies to problems with general trajectory constraints. Under suitable assumptions, we will show that if the base heuristic produces a feasible solution, the rollout algorithm has a cost improvement property: it produces a feasible solution, whose cost is no worse than the base heuristic's cost.

Before going into formal descriptions of the constrained DP problem formulation and the corresponding algorithms, it is worth to revisit the broad outline of the rollout algorithm for deterministic DP:

(a) It constructs a sequence $\{T_0, T_1, \ldots, T_N\}$ of complete system trajectories with monotonically nonincreasing cost (assuming a sequential improvement condition).

(b) The initial trajectory T_0 is the one generated by the base heuristic starting from x_0, and the final trajectory T_N is the one generated by the rollout algorithm.

(c) For each k, the trajectories $T_k, T_{k+1}, \ldots, T_N$ share the same initial portion $(x_0, \tilde{u}_0, \ldots, \tilde{u}_{k-1}, \tilde{x}_k)$.

(d) For each k, the base heuristic is used to generate a number of candidate trajectories, all of which share the initial portion with T_k, up to state $\tilde{x}_k$. These candidate trajectories correspond to the controls $u_k \in U_k(x_k)$. (In the case of fortified rollout, these trajectories include the current "tentative best" trajectory.)

(e) For each k, the next trajectory T_{k+1} is the candidate trajectory that is best in terms of total cost.

In our constrained DP formulation, to be described shortly, we introduce a trajectory constraint $T \in C$, where C is some subset of admissible trajectories. A consequence of this is that some of the candidate trajectories in (d) above, may be infeasible. Our modification to deal with this situation is simple: *we discard all the candidate trajectories that violate the constraint, and we choose T_{k+1} to be the best of the remaining candidate trajectories, the ones that are feasible.*

Of course, for this modification to be viable, we have to guarantee that at least one of the candidate trajectories will satisfy the constraint for every k. For this we will rely on a sequential improvement condition that we will introduce shortly. For the case where this condition does not hold, we will introduce a fortified version of the algorithm, which requires only that the base heuristic generates a feasible trajectory T_0 starting from the initial condition x_0. Thus *to apply reliably the constrained rollout algorithm, we only need to know a single feasible solution*, i.e., a trajectory T_0 that starts at x_0 and satisfies the constraint $T_0 \in C$.

Constrained Problem Formulation

We assume that the state x_k takes values in some (possibly infinite) set and the control u_k takes values in some finite set. The finiteness of the control space is only needed for implementation purposes of the rollout algorithms to be described shortly; simplified versions of the algorithm do not require the finiteness condition. A sequence of the form

$$T = (x_0, u_0, x_1, u_1, \ldots, u_{N-1}, x_N), \qquad (6.26)$$

where

$$x_{k+1} = f_k(x_k, u_k), \qquad k = 0, 1, \ldots, N-1, \qquad (6.27)$$

is referred to as a *complete trajectory*. Our problem is stated succinctly as

$$\min_{T \in C} G(T), \qquad (6.28)$$

where G is some cost function and C is the constraint set.

Note that G need not have the additive form

$$G(T) = g_N(x_N) + \sum_{k=0}^{N-1} g_k(x_k, u_k), \qquad (6.29)$$

Sec. 6.6   Constrained Forms of Rollout Algorithms   **161**

which we have assumed so far. Thus, except for the finiteness of the control space, which is needed for implementation of rollout, this is a very general optimization problem. In fact, later we will simplify the problem further by eliminating the state transition structure of Eq. (6.27).†

Trajectory constraints can arise in a number of ways. A relatively simple example is the standard problem formulation for deterministic DP: an additive cost of the form (6.29), where the controls satisfy the time-uncoupled constraints $u_k \in U_k(x_k)$ [so here C is the set of trajectories that are generated by the system equation with controls satisfying $u_k \in U_k(x_k)$]. In a more complicated constrained DP problem, there may be constraints that couple the controls of different stages such as

$$g_N^m(x_N) + \sum_{k=0}^{N-1} g_k^m(x_k, u_k) \le b^m, \qquad m = 1, \ldots, M, \qquad (6.30)$$

where g_k^m and b^m are given functions and scalars, respectively. An example where difficult trajectory constraints arise is when the control contains some discrete components, which once chosen must remain fixed for multiple time periods.

Here is another discrete optimization example involving the traveling salesman problem.

Example 6.6.1 (A Constrained Form of the Traveling Salesman Problem)

Let us consider a constrained version of the traveling salesman problem of Example 6.2.1. We want to find a minimum travel cost tour that additionally satisfies a safety constraint that the "safety cost" of the tour should be less than a certain threshold; see Fig. 6.6.1. This constraint need not have the additive structure of Eq. (6.30). We are simply given a safety cost for each tour (see the table at the bottom right), which is calculated in a way that is of no further concern to us. In this example, for a tour to be admissible, its safety cost must be less than or equal to 10. Note that the (unconstrained) minimum cost tour, ABDCA, does not satisfy the safety constraint.

Transforming Constrained DP Problems to Unconstrained Problems

Generally, a constrained DP problem can be transformed to an unconstrained DP problem, at the expense of a complicated reformulation of the

---

† Actually, similar to our discussion on model-free rollout in Section 6.4, it is not essential that we know the explicit form of the cost function G and the constraint set C. For our constrained rollout algorithms, it is sufficient to have access to a human or software expert that can determine whether a given trajectory T is feasible, i.e., satisfies the constraint $T \in C$, and also to be able to compare any two feasible trajectories T_1 and T_2, based on some internal process that is unknown to us, without assigning numerical values to them.

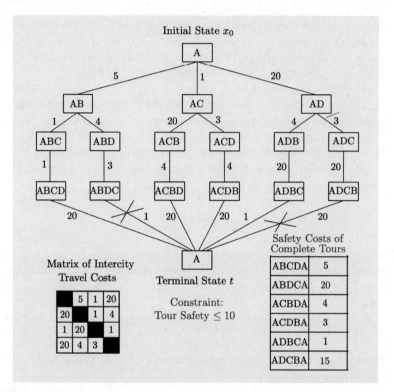

Figure 6.6.1 An example of a constrained traveling salesman problem; cf. Example 6.6.1. We want to find a minimum cost tour that has safety cost less or equal to 10. The safety costs of the six possible tours are given in the table on the right. The (unconstrained) minimum cost tour, ABDCA, does not satisfy the safety constraint. The optimal constrained tour is ABCDA.

state and the system equation. The idea is to redefine the state at stage k to be the partial trajectory

$$y_k = (x_0, u_0, x_1, \ldots, u_{k-1}, x_k),$$

which evolves according to a redefined system equation:

$$y_{k+1} = \big(y_k, u_k, f_k(x_k, u_k)\big).$$

The problem then becomes to find a control sequence that minimizes the terminal cost $G(y_N)$ subject to the constraint $y_N \in C$. This is a problem to which the standard form of DP applies:

$$J_k^*(y_k) = \min_{u_k \in U_k(y_k)} J_{k+1}^*\big(y_k, u_k, f_k(x_k, u_k)\big), \qquad k = 0, \ldots, N-1,$$

where

$$J_N^*(y_N) = G(y_N),$$

Sec. 6.6 Constrained Forms of Rollout Algorithms

and for $k = 0, \ldots, N-1$, the constraint set $U_k(y_k)$ is the subset of controls for which it is possible to attain feasibility. Thus $U_k(y_k)$ is the set of u_k such that there exist $u_{k+1}, \ldots, u_{N-1}$ and corresponding $x_{k+1}, \ldots, x_N$, which together with y_k, satisfy

$$(y_k, u_k, x_{k+1}, u_{k+1}, \ldots, x_{N-1}, u_{N-1}, x_N) \in C.$$

The reformulation to an unconstrained problem just described is typically impractical, because the associated computation can be overwhelming. However, it provides guidance for structuring a constrained rollout algorithm, which we describe next. Moreover, it allows the interpretation of this constrained rollout algorithm in terms of the Newton step, which is the central theme of this monograph.

Using a Base Heuristic for Constrained Rollout

We will now describe formally the constrained rollout algorithm. We assume the availability of a base heuristic, which for any given partial trajectory

$$y_k = (x_0, u_0, x_1, \ldots, u_{k-1}, x_k),$$

can produce a (complementary) partial trajectory

$$R(y_k) = (x_k, u_k, x_{k+1}, u_{k+1}, \ldots, u_{N-1}, x_N),$$

that starts at x_k and satisfies the system equation

$$x_{t+1} = f_t(x_t, u_t), \qquad t = k, \ldots, N-1.$$

Thus, given y_k and any control u_k, we can use the base heuristic to obtain a complete trajectory as follows:

(a) Generate the next state $x_{k+1} = f_k(x_k, u_k)$.

(b) Extend y_k to obtain the partial trajectory

$$y_{k+1} = \big(y_k, u_k, f_k(x_k, u_k)\big).$$

(c) Run the base heuristic from y_{k+1} to obtain the partial trajectory $R(y_{k+1})$.

(d) Join the two partial trajectories y_{k+1} and $R(y_{k+1})$ to obtain the complete trajectory $\big(y_k, u_k, R(y_{k+1})\big)$, which is denoted by $T_k(y_k, u_k)$:

$$T_k(y_k, u_k) = \big(y_k, u_k, R(y_{k+1})\big). \tag{6.31}$$

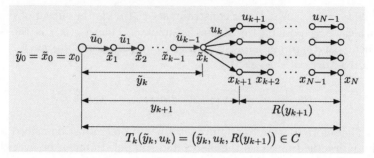

Figure 6.6.2 The trajectory generation mechanism of the rollout algorithm. At stage k, and given the current partial trajectory

$$\tilde{y}_k = (\tilde{x}_0, \tilde{u}_0, \tilde{x}_1, \ldots, \tilde{u}_{k-1}, \tilde{x}_k),$$

which starts at $\tilde{x}_0$ and ends at $\tilde{x}_k$, we consider all possible next states $x_{k+1} = f_k(\tilde{x}_k, u_k)$, run the base heuristic starting at $y_{k+1} = (\tilde{y}_k, u_k, x_{k+1})$, and form the complete trajectory $T_k(\tilde{y}_k, u_k)$. Then the rollout algorithm:

(a) Finds $\tilde{u}_k$, the control that minimizes the cost $G(T_k(\tilde{y}_k, u_k))$ over all u_k for which the complete trajectory $T_k(\tilde{y}_k, u_k)$ is feasible.

(b) Extends $\tilde{y}_k$ by $(\tilde{u}_k, f_k(\tilde{x}_k, \tilde{u}_k))$ to form $\tilde{y}_{k+1}$.

This process is illustrated in Fig. 6.6.2. Note that the partial trajectory $R(y_{k+1})$ produced by the base heuristic depends on the entire partial trajectory y_{k+1}, not just the state x_{k+1}.

A complete trajectory $T_k(y_k, u_k)$ of the form (6.31) is generally feasible for only the subset $U_k(y_k)$ of controls u_k that maintain feasibility:

$$U_k(y_k) = \{u_k \mid T_k(y_k, u_k) \in C\}. \tag{6.32}$$

Our rollout algorithm starts from a given initial state $\tilde{y}_0 = \tilde{x}_0$, and generates successive partial trajectories $\tilde{y}_1, \ldots, \tilde{y}_N$, of the form

$$\tilde{y}_{k+1} = (\tilde{y}_k, \tilde{u}_k, f_k(\tilde{x}_k, \tilde{u}_k)), \qquad k = 0, \ldots, N-1, \tag{6.33}$$

where $\tilde{x}_k$ is the last state component of $\tilde{y}_k$, and $\tilde{u}_k$ is a control that minimizes the heuristic cost $G(T_k(\tilde{y}_k, u_k))$ over all u_k for which $T_k(\tilde{y}_k, u_k)$ is feasible. Thus at stage k, the algorithm forms the set $U_k(\tilde{y}_k)$ [cf. Eq. (6.32)] and selects from $U_k(\tilde{y}_k)$ a control $\tilde{u}_k$ that minimizes the cost of the complete trajectory $T_k(\tilde{y}_k, u_k)$:

$$\tilde{u}_k \in \arg\min_{u_k \in U_k(\tilde{y}_k)} G(T_k(\tilde{y}_k, u_k)); \tag{6.34}$$

see Fig. 6.6.2. The objective is to produce a feasible final complete trajectory $\tilde{y}_N$, which has a cost $G(\tilde{y}_N)$ that is no larger than the cost of $R(\tilde{y}_0)$ produced by the base heuristic starting from $\tilde{y}_0$, i.e.,

$$G(\tilde{y}_N) \leq G(R(\tilde{y}_0)).$$

Sec. 6.6   Constrained Forms of Rollout Algorithms   **165**

Note that $T_k(\tilde{y}_k, u_k)$ is not guaranteed to be feasible for any given u_k (i.e., may not belong to C), but we will assume that the constraint set $U_k(\tilde{y}_k)$ of problem (6.34) is nonempty, so that our rollout algorithm is well-defined. We will later modify our algorithm so that it is well-defined under the weaker assumption that just *the complete trajectory generated by the base heuristic starting from the initial state $\tilde{y}_0$ is feasible*, i.e., $R(\tilde{y}_0) \in C$.

Constrained Rollout Algorithm

The algorithm starts at stage 0 and sequentially proceeds to the last stage. At the typical stage k, it has constructed a partial trajectory

$$\tilde{y}_k = (\tilde{x}_0, \tilde{u}_0, \tilde{x}_1, \ldots, \tilde{u}_{k-1}, \tilde{x}_k) \qquad (6.35)$$

that starts at the given initial state $\tilde{y}_0 = \tilde{x}_0$, and is such that

$$\tilde{x}_{t+1} = f_t(\tilde{x}_t, \tilde{u}_t), \qquad t = 0, 1, \ldots, k-1.$$

The algorithm then forms the set of controls

$$U_k(\tilde{y}_k) = \{u_k \mid T_k(\tilde{y}_k, u_k) \in C\}$$

that is consistent with feasibility [cf. Eq. (6.32)], and chooses a control $\tilde{u}_k \in U_k(\tilde{y}_k)$ according to the minimization

$$\tilde{u}_k \in \arg\min_{u_k \in U_k(\tilde{y}_k)} G(T_k(\tilde{y}_k, u_k)), \qquad (6.36)$$

[cf. Eq. (6.34)], where

$$T_k(\tilde{y}_k, u_k) = \Big(\tilde{y}_k, u_k, R\big(\tilde{y}_k, u_k, f_k(\tilde{x}_k, u_k)\big)\Big);$$

[cf. Eq. (6.31)]. Finally, the algorithm sets

$$\tilde{x}_{k+1} = f_k(\tilde{x}_k, \tilde{u}_k), \qquad \tilde{y}_{k+1} = (\tilde{y}_k, \tilde{u}_k, \tilde{x}_{k+1}),$$

[cf. Eq. (6.33)], thus obtaining the partial trajectory $\tilde{y}_{k+1}$ to start the next stage.

It can be seen that our constrained rollout algorithm is not much more complicated or computationally demanding than its unconstrained version where the constraint $T \in C$ is not present (as long as checking feasibility of a complete trajectory T is not computationally demanding). Note, however, that our algorithm makes essential use of the deterministic character of the problem, and does not admit a straightforward extension to stochastic problems, since checking feasibility of a complete trajectory is typically difficult in the context of these problems.

The rollout algorithm just described is illustrated in Fig. 6.6.3 for our earlier traveling salesman Example 6.6.1. Here we want to find a minimum travel cost tour that additionally satisfies a safety constraint, namely that the "safety cost" of the tour should be less than a certain threshold. Note that the minimum cost tour, ABDCA, in this example does not satisfy the safety constraint. Moreover, the tour ABCDA obtained by the rollout algorithm has barely smaller cost than the tour ACDBA generated by the base heuristic starting from A. In fact if the travel cost D→A were larger, say 25, the tour produced by constrained rollout would be more costly than the one produced by the base heuristic starting from A. This points to the need for a constrained version of the notion of sequential improvement and for a fortified variant of the algorithm, which we discuss next.

Sequential Consistency, Sequential Improvement, and the Cost Improvement Property

We will now introduce sequential consistency and sequential improvement conditions guaranteeing that the control set $U_k(\tilde{y}_k)$ in the minimization (6.36) is nonempty, and that the costs of the complete trajectories $T_k(\tilde{y}_k, \tilde{u}_k)$ are improving with each k in the sense that

$$G\big(T_{k+1}(\tilde{y}_{k+1}, \tilde{u}_{k+1})\big) \leq G\big(T_k(\tilde{y}_k, \tilde{u}_k)\big), \qquad k = 0, 1, \ldots, N-1,$$

while at the first step of the algorithm we have

$$G\big(T_0(\tilde{y}_0, \tilde{u}_0)\big) \leq G\big(R(\tilde{y}_0)\big).$$

It will then follow that the cost improvement property

$$G(\tilde{y}_N) \leq G\big(R(\tilde{y}_0)\big)$$

holds.

Definition 6.6.1: We say that the base heuristic is *sequentially consistent* if whenever it generates a partial trajectory

$$(x_k, u_k, x_{k+1}, u_{k+1}, \ldots, u_{N-1}, x_N),$$

starting from a partial trajectory y_k, it also generates the partial trajectory

$$(x_{k+1}, u_{k+1}, x_{k+2}, u_{k+2}, \ldots, u_{N-1}, x_N),$$

starting from the partial trajectory $y_{k+1} = (y_k, u_k, x_{k+1})$.

Sec. 6.6   Constrained Forms of Rollout Algorithms   167

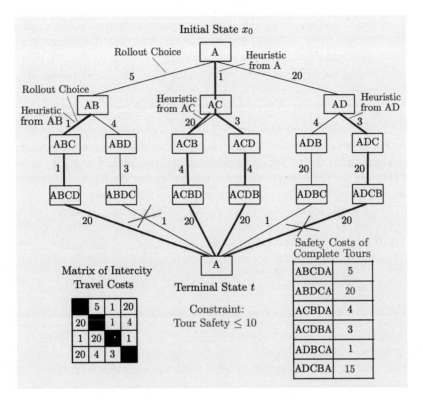

Figure 6.6.3 The constrained traveling salesman problem; cf. Example 6.6.1, and its rollout solution using the base heuristic shown, which completes a partial tour as follows:

    At A it yields ACDBA.
    At AB it yields ABCDA.
    At AC it yields ACBDA.
    At AD it yields ADCBA.

This base heuristic is not assumed to have any special structure. It is just capable of completing every partial tour without regard to any additional considerations. Thus for example the heuristic generates at A the complete tour ACDBA, and it switches to the tour ACBDA once the salesman moves to AC.

At city A, the rollout algorithm:

(a) Considers the partial tours AB, AC, and AD.

(b) Uses the base heuristic to obtain the corresponding complete tours ABCDA, ACBDA, and ADCBA.

(c) Discards ADCBA as being infeasible.

(d) Compares the other two tours, ABCDA and ACBDA, finds ABCDA to have smaller cost, and selects the partial tour AB.

(e) At AB, it considers the partial tours ABC and ABD.

(f) It uses the base heuristic to obtain the corresponding complete tours ABCDA and ABDCA, and discards ABDCA as being infeasible.

(g) It finally selects the complete tour ABCDA.

As we have noted in the context of unconstrained rollout, greedy heuristics tend to be sequentially consistent. Also any policy [a sequence of feedback control functions $\mu_k(y_k)$, $k = 0, 1, \ldots, N-1$] for the DP problem of minimizing the terminal cost $G(y_N)$ subject to the system equation

$$y_{k+1} = \big(y_k, u_k, f_k(x_k, u_k)\big)$$

and the feasibility constraint $y_N \in C$ can be seen to be sequentially consistent. For an example where sequential consistency is violated, consider the base heuristic of the traveling salesman Example 6.6.1. From Fig. 6.6.3, it can be seen that the base heuristic at A generates ACDBA, but from AC it generates ACBDA, thus violating sequential consistency.

For a given partial trajectory y_k, let us denote by $y_k \cup R(y_k)$ the complete trajectory obtained by joining y_k with the partial trajectory generated by the base heuristic starting from y_k. Thus if

$$y_k = (x_0, u_0, \ldots, u_{k-1}, x_k)$$

and

$$R(y_k) = (x_k, u_{k+1}, \ldots, u_{N-1}, x_N),$$

we have

$$y_k \cup R(y_k) = (x_0, u_0, \ldots, u_{k-1}, x_k, u_{k+1}, \ldots, u_{N-1}, x_N).$$

Definition 6.6.2: We say that the base heuristic is *sequentially improving* if for every $k = 0, 1, \ldots, N-1$ and partial trajectory y_k for which $y_k \cup R(y_k) \in C$, the set $U_k(y_k)$ is nonempty, and we have

$$G\big(y_k \cup R(y_k)\big) \geq \min_{u_k \in U_k(y_k)} G\big(T_k(y_k, u_k)\big). \tag{6.37}$$

Note that for a base heuristic that is not sequentially consistent, the condition $y_k \cup R(y_k) \in C$ does not imply that the set $U_k(y_k)$ is nonempty. The reason is that starting from $\big(y_k, u_k, f_k(x_k, u_k)\big)$ the base heuristic may generate a different trajectory than from y_k, even if it applies u_k at y_k. Thus we need to include nonemptiness of $U_k(y_k)$ as a requirement in the preceding definition of sequential improvement (in the fortified version of the algorithm to be discussed shortly, this requirement will be removed).

On the other hand, if the base heuristic is sequentially consistent, it is also sequentially improving. The reason is that for a sequentially consistent heuristic, $y_k \cup R(y_k)$ is equal to one of the trajectories contained in the set

$$\big\{T_k(y_k, u_k) \mid u_k \in U_k(y_k)\big\}.$$

Sec. 6.6   Constrained Forms of Rollout Algorithms

Our main result is contained in the following proposition.

Proposition 6.6.1: (Cost Improvement for Constrained Rollout) Assume that the base heuristic is sequentially improving and generates a feasible complete trajectory starting from the initial state $\tilde{y}_0 = \tilde{x}_0$, i.e., $R(\tilde{y}_0) \in C$. Then for each k, the set $U_k(\tilde{y}_k)$ is nonempty, and we have

$$G\big(R(\tilde{y}_0)\big) \geq G\big(T_0(\tilde{y}_0, \tilde{u}_0)\big)$$
$$\geq G\big(T_1(\tilde{y}_1, \tilde{u}_1)\big)$$
$$\geq \cdots$$
$$\geq G\big(T_{N-1}(\tilde{y}_{N-1}, \tilde{u}_{N-1})\big)$$
$$= G(\tilde{y}_N),$$

where

$$T_k(\tilde{y}_k, \tilde{u}_k) = \big(\tilde{y}_k, \tilde{u}_k, R(\tilde{y}_{k+1})\big);$$

cf. Eq. (6.31). In particular, the final trajectory $\tilde{y}_N$ generated by the constrained rollout algorithm is feasible and has no larger cost than the trajectory $R(\tilde{y}_0)$ generated by the base heuristic starting from the initial state.

Proof: Consider $R(\tilde{y}_0)$, the complete trajectory generated by the base heuristic starting from $\tilde{y}_0$. Since $\tilde{y}_0 \cup R(\tilde{y}_0) = R(\tilde{y}_0) \in C$ by assumption, it follows from the sequential improvement definition, that the set $U_0(\tilde{y}_0)$ is nonempty and we have

$$G\big(R(\tilde{y}_0)\big) \geq G\big(T_0(\tilde{y}_0, \tilde{u}_0)\big),$$

[cf. Eq. (6.37)], while $T_0(\tilde{y}_0, \tilde{u}_0) \in C$.

The preceding argument can be repeated for the next stage, by replacing $\tilde{y}_0$ with $\tilde{y}_1$, and $R(\tilde{y}_0)$ with $T_0(\tilde{y}_0, \tilde{u}_0)$. Since $\tilde{y}_1 \cup R(\tilde{y}_1) = T_0(\tilde{y}_0, \tilde{u}_0) \in C$, from the sequential improvement definition, the set $U_1(\tilde{y}_1)$ is nonempty and we have

$$G\big(T_0(\tilde{y}_0, \tilde{u}_0)\big) = G\big(\tilde{y}_1 \cup R(\tilde{y}_1)\big) \geq G\big(T_1(\tilde{y}_1, \tilde{u}_1)\big),$$

[cf. Eq. (6.37)], while $T_1(\tilde{y}_1, \tilde{u}_1) \in C$. Similarly, the argument can be successively repeated for every k, to verify that $U_k(\tilde{y}_k)$ is nonempty and that $G\big(T_k(\tilde{y}_k, \tilde{u}_k)\big) \geq G\big(T_{k+1}(\tilde{y}_{k+1}, \tilde{u}_{k+1})\big)$ for all k.   **Q.E.D.**

Proposition 6.6.1 establishes the fundamental cost improvement property for constrained rollout under the sequential improvement condition. On the other hand we may construct examples where the sequential improvement condition (6.37) is violated and the cost of the solution produced by rollout is larger than the cost of the solution produced by the base heuristic starting from the initial state (cf. the unconstrained rollout Example 6.4.2).

In the case of the traveling salesman Example 6.6.1, it can be verified that the base heuristic specified in Fig. 6.6.3 is sequentially improving. However, if the travel cost D→A were larger, say 25, then it can be verified that the definition of sequential improvement would be violated at A, and the tour produced by constrained rollout would be more costly than the one produced by the base heuristic starting from A.

The Fortified Rollout Algorithm and Other Variations

We will now discuss some variations and extensions of the constrained rollout algorithm. Let us first consider the case where the sequential improvement assumption is not satisfied. Then it may happen that given the current partial trajectory $\tilde{y}_k$, the set of controls $U_k(\tilde{y}_k)$ that corresponds to feasible trajectories $T_k(\tilde{y}_k, u_k)$ [cf. Eq. (6.32)] is empty, in which case the rollout algorithm cannot extend the partial trajectory $\tilde{y}_k$ further. To bypass this difficulty, we introduce a *fortified constrained rollout algorithm*, patterned after the fortified algorithm given earlier. For validity of this algorithm, *we require that the base heuristic generates a feasible complete trajectory $R(\tilde{y}_0)$ starting from the initial state $\tilde{y}_0$.*

The fortified constrained rollout algorithm, in addition to the current partial trajectory

$$\tilde{y}_k = (\tilde{x}_0, \tilde{u}_0, \tilde{x}_1, \ldots, \tilde{u}_{k-1}, \tilde{x}_k),$$

maintains a complete trajectory $\hat{T}_k$, called *tentative best trajectory*, which is feasible (i.e., $\hat{T}_k \in C$) and agrees with $\tilde{y}_k$ up to state $\tilde{x}_k$, i.e., $\hat{T}_k$ has the form

$$\hat{T}_k = (\tilde{x}_0, \tilde{u}_0, \tilde{x}_1, \ldots, \tilde{u}_{k-1}, \tilde{x}_k, \overline{u}_k, \overline{x}_{k+1}, \ldots, \overline{u}_{N-1}, \overline{x}_N), \qquad (6.38)$$

for some $\overline{u}_k, \overline{x}_{k+1}, \ldots, \overline{u}_{N-1}, \overline{x}_N$ such that

$$\overline{x}_{k+1} = f_k(\tilde{x}_k, \overline{u}_k), \qquad \overline{x}_{t+1} = f_t(\overline{x}_t, \overline{u}_t), \qquad t = k+1, \ldots, N-1.$$

Initially, $\hat{T}_0$ is the complete trajectory $R(\tilde{y}_0)$, generated by the base heuristic starting from $\tilde{y}_0$, which is assumed to be feasible. At stage k, the algorithm forms the subset $\hat{U}_k(\tilde{y}_k)$ of controls $u_k \in U_k(\tilde{y}_k)$ such that the corresponding $T_k(\tilde{y}_k, u_k)$ is not only feasible, but also has cost that is no larger than the one of the current tentative best trajectory:

$$\hat{U}_k(\tilde{y}_k) = \left\{ u_k \in U_k(\tilde{y}_k) \mid G\big(T_k(\tilde{y}_k, u_k)\big) \leq G(\hat{T}_k) \right\}.$$

There are two cases to consider at state k:

(1) *The set $\hat{U}_k(\tilde{y}_k)$ is nonempty.* Then the algorithm forms the partial trajectory $\tilde{y}_{k+1} = (\tilde{y}_k, \tilde{u}_k, \tilde{x}_{k+1})$, where

$$\tilde{u}_k \in \arg\min_{u_k \in \hat{U}_k(\tilde{y}_k)} G\big(T_k(\tilde{y}_k, u_k)\big), \qquad \tilde{x}_{k+1} = f_k(\tilde{x}_k, \tilde{u}_k),$$

and sets $T_k(\tilde{y}_k, \tilde{u}_k)$ as the new tentative best trajectory, i.e.,

$$\hat{T}_{k+1} = T_k(\tilde{y}_k, \tilde{u}_k).$$

(2) *The set $\hat{U}_k(\tilde{y}_k)$ is empty.* Then, the algorithm forms the partial trajectory $\tilde{y}_{k+1} = (\tilde{y}_k, \tilde{u}_k, \tilde{x}_{k+1})$, where

$$\tilde{u}_k = \overline{u}_k, \qquad \tilde{x}_{k+1} = \overline{x}_{k+1},$$

and $\overline{u}_k, \overline{x}_{k+1}$ are the control and state subsequent to $\tilde{x}_k$ in the current tentative best trajectory $\hat{T}_k$ [cf. Eq. (6.38)], and leaves $\hat{T}_k$ unchanged, i.e.,

$$\hat{T}_{k+1} = \hat{T}_k.$$

It can be seen that the fortified constrained rollout algorithm will follow the initial complete trajectory $\hat{T}_0$, the one generated by the base heuristic starting from $\tilde{y}_0$, up to a stage k where it will discover a new feasible complete trajectory with smaller cost to replace $\hat{T}_0$ as the tentative best trajectory. Similarly, the new tentative best trajectory $\hat{T}_k$ may be subsequently replaced by another feasible trajectory with smaller cost, etc.

Note that if the base heuristic is sequentially improving, and the fortified rollout algorithm will generate the same complete trajectory as the (nonfortified) rollout algorithm given earlier, with the tentative best trajectory $\hat{T}_{k+1}$ being equal to the complete trajectory $T_k(\tilde{y}_k, \tilde{u}_k)$ for all k. The reason is that if the base heuristic is sequentially improving, the controls $\tilde{u}_k$ generated by the nonfortified algorithm belong to the set $\hat{U}_k(\tilde{y}_k)$ [by Prop. 6.6.1, case (1) above will hold].

However, it can be verified that even when the base heuristic is not sequentially improving, the fortified rollout algorithm will generate a complete trajectory that is feasible and has cost that is no worse than the cost of the complete trajectory generated by the base heuristic starting from $\tilde{y}_0$. This is because each tentative best trajectory has a cost that is no worse than the one of its predecessor, and the initial tentative best trajectory is just the trajectory generated by the base heuristic starting from the initial condition $\tilde{y}_0$.

Tree-Based Rollout Algorithms

It is possible to improve the performance of the rollout algorithm at the expense of maintaining more than one partial trajectory. In particular, instead of the partial trajectory $\tilde{y}_k$ of Eq. (6.35), we can maintain a *tree* of partial trajectories that is rooted at $\tilde{y}_0$. These trajectories need not have equal length, i.e., they need not involve the same number of stages. At each step of the algorithm, we select a single partial trajectory from this tree, and execute the rollout algorithm's step as if this partial trajectory were the only one. Let this partial trajectory have k stages and denote it by $\tilde{y}_k$. Then we extend $\tilde{y}_k$ similar to our earlier rollout algorithm, with possibly multiple feasible trajectories. There is also a fortified version of this algorithm where a tentative best trajectory is maintained, which is the minimum cost complete trajectory generated thus far.

The aim of the tree-based algorithm is to obtain improved performance, essentially because it can go back and extend partial trajectories that were generated and temporarily abandoned at previous stages. The net result is a more flexible algorithm that is capable of examining more alternative trajectories. Note also that there is considerable freedom to select the number of partial trajectories maintained in the tree.

We finally mention a drawback of the tree-based algorithm: it is suitable for off-line computation, but it cannot be applied in an on-line context, where the rollout control selection is made after the current state becomes known as the system evolves in real-time.

Constrained Multiagent Rollout

Let us consider a special structure of the control space, where the control u_k consists of m components, $u_k = (u_k^1, \ldots, u_k^m)$, each belonging to a corresponding set $U_k^\ell(x_k)$, $\ell = 1, \ldots, m$. Thus the control space at stage k is the Cartesian product

$$U_k(x_k) = U_k^1(x_k) \times \cdots \times U_k^m(x_k).$$

We refer to this as the *multiagent case*, motivated by the special case where each component u_k^ℓ, $\ell = 1, \ldots, m$, is chosen by a separate agent ℓ at stage k.

Similar to the stochastic unconstrained case of Chapter 3, we can introduce a modified but equivalent problem, involving one-at-a-time agent control selection. In particular, at the generic state x_k, we break down the control u_k into the sequence of the m controls $u_k^1, u_k^2, \ldots, u_k^m$, and between x_k and the next state $x_{k+1} = f_k(x_k, u_k)$, we introduce artificial intermediate "states"

$$(x_k, u_k^1), (x_k, u_k^1, u_k^2), \ldots, (x_k, u_k^1, \ldots, u_k^{m-1}),$$

and corresponding transitions. The choice of the last control component u_k^m at "state" $(x_k, u_k^1, \ldots, u_k^{m-1})$ marks the transition at cost $g_k(x_k, u_k)$ to the next state $x_{k+1} = f_k(x_k, u_k)$ according to the system equation. It is

Sec. 6.7   Adaptive Control by Rollout with a POMDP Formulation   173

evident that this reformulated problem is equivalent to the original, since any control choice that is possible in one problem is also possible in the other problem, with the same cost.

By working with the reformulated problem, we can consider a rollout algorithm that requires a sequence of m minimizations per stage, one over each of the control components $u_k^1, \ldots, u_k^m$, with the past controls already determined by the rollout algorithm, and the future controls determined by running the base heuristic. Assuming a maximum of n elements in the control component spaces $U_k^\ell(x_k)$, $\ell = 1, \ldots, m$, the computation required for the m single control component minimizations is of order $O(nm)$ per stage. By contrast the standard rollout minimization (6.36) involves the computation and comparison of as many as n^m terms $G\bigl(T_k(\tilde{y}_k, u_k)\bigr)$ per stage.

6.7 ADAPTIVE CONTROL BY ROLLOUT WITH A POMDP FORMULATION

In this section, we discuss various approaches for the approximate solution of Partially Observed Markovian Decision Problems (POMDP) with a special structure, which is well-suited for adaptive control, as well as other contexts that involve search for a hidden object. It is well known that POMDP are among the most challenging DP problems, and nearly always require the use of approximations for (suboptimal) solution.

The application and implementation of rollout and approximate PI methods to general finite-state POMDP is described in the author's RL book [Ber19a] (Section 5.7.3). Here we will focus attention on a special class of POMDP where the state consists of two components:

(a) A perfectly observed component x_k that evolves over time according to a discrete-time equation.

(b) A component θ which is unobserved but stays constant, and is estimated through the perfect observations of the component x_k.

We view θ as a parameter in the system equation that governs the evolution of x_k. Thus we have

$$x_{k+1} = f_k(x_k, \theta, u_k, w_k), \tag{6.39}$$

where u_k is the control at time k, selected from a set $U_k(x_k)$, and w_k is a random disturbance with given probability distribution that depends on (x_k, θ, u_k). We will assume that θ can take one of m known values $\theta^1, \ldots, \theta^m$:

$$\theta \in \{\theta^1, \ldots, \theta^m\}.$$

The a priori probability distribution of θ is given and is updated based on the observed values of the state components x_k and the applied controls

u_k. In particular, we assume that the information vector

$$I_k = \{x_0, \ldots, x_k, u_0, \ldots, u_{k-1}\}$$

is available at time k, and is used to compute the conditional probabilities

$$b_{k,i} = P\{\theta = \theta^i \mid I_k\}, \qquad i = 1, \ldots, m.$$

These probabilities form a vector

$$b_k = (b_{k,1}, \ldots, b_{k,m}),$$

which together with the perfectly observed state x_k, form the pair (x_k, b_k) that is commonly called the *belief state* of the POMDP at time k.

Note that according to the classical methodology of POMDP (see e.g., [Ber17a], Chapter 4), the belief component b_{k+1} is determined by the belief state (x_k, b_k), the control u_k, and the observation obtained at time $k+1$, i.e., x_{k+1}. Thus b_k can be updated according to an equation of the form

$$b_{k+1} = B_k(x_k, b_k, u_k, x_{k+1}),$$

where B_k is an appropriate function, which can be viewed as a recursive estimator of θ. There are several approaches to implement this estimator (perhaps with some approximation error), including the use of Bayes' rule and the simulation-based method of particle filtering.

The preceding mathematical model forms the basis for a classical adaptive control formulation, where each θ^i represents an unknown set of system parameters, and the computation of the belief probabilities $b_{k,i}$ can be viewed as the outcome of a system identification algorithm. In this context, the problem becomes one of *dual control*, a combined identification and control problem, whose optimal solution is notoriously difficult.

Another interesting context arises in search problems, where θ specifies the locations of one or more objects of interest within a given space. Some puzzles, including the popular Wordle game, fall within this category, as we will discuss briefly later in this section.

The Exact DP Algorithm - Approximation in Value Space

We will now describe an exact DP algorithm that operates in the space of information vectors I_k. To describe this algorithm, let us denote by $J_k(I_k)$ the optimal cost starting at information vector I_k at time k. Using the equation

$$I_{k+1} = (I_k, x_{k+1}, u_k) = \big(I_k, f_k(x_k, \theta, u_k, w_k), u_k\big),$$

Sec. 6.7   Adaptive Control by Rollout with a POMDP Formulation   175

the algorithm takes the form

$$J_k(I_k) = \min_{u_k \in U_k(x_k)} E_{\theta, w_k} \Big\{ g_k(x_k, \theta, u_k, w_k) + J_{k+1}\big(I_k, f_k(x_k, \theta, u_k, w_k), u_k\big) \mid I_k, u_k \Big\},$$
(6.40)

for $k = 0, \ldots, N-1$, with $J_N(I_N) = g_N(x_N)$; see e.g., the DP textbook [Ber17a], Section 4.1.

By using the law of iterated expectations,

$$E_{\theta, w_k}\{\cdot \mid I_k, u_k\} = E_\theta\{E_{w_k}\{\cdot \mid I_k, \theta, u_k\} \mid I_k, u_k\},$$

we can rewrite this DP algorithm as

$$J_k(I_k) = \min_{u_k \in U_k(x_k)} \sum_{i=1}^m b_{k,i} E_{w_k} \Big\{ g_k(x_k, \theta^i, u_k, w_k) + J_{k+1}\big(I_k, f_k(x_k, \theta^i, u_k, w_k), u_k\big) \mid I_k, \theta^i, u_k \Big\}.$$
(6.41)

The summation over i above represents the expected value of θ conditioned on I_k and u_k.

The algorithm (6.41) is typically very hard to implement, because of the dependence of J_{k+1} on the entire information vector I_{k+1}, which expands in size according to

$$I_{k+1} = (I_k, x_{k+1}, u_k).$$

To address this implementation difficulty, we may use approximation in value space, based on replacing J_{k+1} in the DP algorithm (6.40) with some function that can be obtained (either off-line or on-line) with a tractable computation.

One approximation possibility is based on the use of the optimal cost function corresponding to each parameter value θ^i,

$$\hat{J}^i_{k+1}(x_{k+1}), \qquad i = 1, \ldots, m. \tag{6.42}$$

Here, $\hat{J}^i_{k+1}(x_{k+1})$ is the optimal cost that would be obtained starting from state x_{k+1} under the assumption that $\theta = \theta^i$; this corresponds to a perfect state information problem. Then an approximation in value space scheme with one-step lookahead minimization is given by

$$\tilde{u}_k \in \arg\min_{u_k \in U_k(x_k)} \sum_{i=1}^m b_{k,i} E_{w_k} \Big\{ g_k(x_k, \theta^i, u_k, w_k) + \hat{J}^i_{k+1}\big(f_k(x_k, \theta^i, u_k, w_k)\big) \mid x_k, \theta^i, u_k \Big\}.$$
(6.43)

In particular, instead of the optimal control, which minimizes the optimal Q-factor of (I_k, u_k) appearing in the right side of Eq. (6.40), we apply control $\tilde{u}_k$ that minimizes the expected value over θ of the optimal Q-factors that correspond to fixed values of θ.

A simpler version of this approach is to use the same function $\hat{J}_{k+1}^i$ for every i. However, the dependence on i may be useful in some contexts where differences in the value of i may have a radical effect on the qualitative character of the system equation.

Generally, the optimal costs $\hat{J}_{k+1}^i(x_{k+1})$ that correspond to the different parameter values θ^i [cf. Eq. (6.42)] may be hard to compute, despite their perfect state information structure.† An alternative possibility is to use off-line trained feature-based or neural network-based approximations to $\hat{J}_{k+1}^i(x_{k+1})$.

In the case where the horizon is infinite, it is reasonable to expect that the estimate of the parameter θ improves over time, and that with a suitable estimation scheme, it converges asymptotically to the correct value of θ, call it θ^*, i.e.,

$$\lim_{k \to \infty} b_{k,i} = \begin{cases} 1 & \text{if } \theta^i = \theta^*, \\ 0 & \text{if } \theta^i \neq \theta^*. \end{cases}$$

Then it can be seen that the generated one-step lookahead controls $\tilde{u}_k$ are asymptotically obtained from the Bellman equation that corresponds to the correct parameter θ^*, and are typically optimal in some asymptotic sense. Schemes of this type have been discussed in the adaptive control literature since the 70s; see e.g., Mandl [Man74], Doshi and Shreve [DoS80], Kumar and Lin [KuL82], Kumar [Kum85]. Moreover, some of the pitfalls of performing parameter identification while simultaneously applying adaptive control have been described by Borkar and Varaiya [BoV79], and by Kumar [Kum83]; see [Ber17a], Section 6.8 for a related discussion.

Rollout

Another possibility is to use the costs of given policies π^i in place of the optimal costs $\hat{J}_{k+1}^i(x_{k+1})$. In this case the one-step lookahead scheme (6.43) takes the form

$$\tilde{u}_k \in \arg \min_{u_k \in U_k(x_k)} \sum_{i=1}^m b_{k,i} E_{w_k} \Big\{ g_k(x_k, \theta^i, u_k, w_k) + \hat{J}_{k+1,\pi^i}^i\big(f_k(x_k, \theta^i, u_k, w_k)\big) \mid x_k, \theta^i, u_k \Big\}, \quad (6.44)$$

---

† In favorable special cases, such as linear quadratic problems, the optimal costs $\hat{J}_{k+1}^i(x_{k+1})$ may be easily calculated in closed form. Still, however, even in such cases the calculation of the belief probabilities $b_{k,i}$ may not be simple, and may require the use of a system identification algorithm.

Sec. 6.7 Adaptive Control by Rollout with a POMDP Formulation

and has the character of a rollout algorithm, with $\pi^i = \{\mu_0^i, \ldots, \mu_{N-1}^i\}$, $i = 1, \ldots, m$, being known base policies, with components μ_k^i that depend only on x_k. Here, the term

$$\hat{J}_{k+1,\pi^i}^i\big(f_k(x_k, \theta^i, u_k, w_k)\big)$$

in Eq. (6.44) is the cost of the base policy π^i, calculated starting from the next state

$$x_{k+1} = f_k(x_k, \theta^i, u_k, w_k),$$

under the assumption that θ will stay fixed at the value $\theta = \theta^i$ until the end of the horizon.

This algorithm is related to the adaptive control/rollout algorithm that we discussed earlier in Section 5.2. Indeed, when the belief probabilities $b_{k,i}$ imply certainty, i.e., $b_{k,\bar{i}} = 1$ for some parameter index $\bar{i}$, and $b_{k,i} = 0$ for $i \neq \bar{i}$, the algorithm (6.44) is identical to the rollout by re-optimization algorithm of Section 5.2, where it is assumed that the model of the system has been estimated exactly. Also if all the policies π^i are the same, a cost improvement property similar to the ones shown earlier can be proved. For further discussion and connections to the Bayesian optimization methodology, we refer to the author's paper [Ber22c].

The Case of a Deterministic System

Let us now consider the case where the system (6.39) is deterministic of the form

$$x_{k+1} = f_k(x_k, \theta, u_k). \tag{6.45}$$

Then, while the problem still has a stochastic character due to the uncertainty about the value of θ, the DP algorithm (6.41) and its approximation in value space counterparts are greatly simplified because there is no expectation over w_k to contend with. Indeed, given a state x_k, a parameter θ^i, and a control u_k, the on-line computation of the control of the rollout-like algorithm (6.44), takes the form

$$\tilde{u}_k \in \arg\min_{u_k \in U_k(x_k)} \sum_{i=1}^{m} b_{k,i}\Big(g_k(x_k, \theta^i, u_k) + \hat{J}_{k+1,\pi^i}^i\big(f_k(x_k, \theta^i, u_k)\big)\Big). \tag{6.46}$$

The computation of $\hat{J}_{k+1,\pi^i}^i\big(f_k(x_k, \theta^i, u_k)\big)$ involves a deterministic propagation from the state x_{k+1} of Eq. (6.45) up to the end of the horizon, using the base policy π^i, while assuming that θ is fixed at the value θ^i.

In particular, the term

$$Q_k(u_k, \theta^i) = g_k(x_k, \theta^i, u_k) + \hat{J}_{k+1,\pi^i}^i\big(f_k(x_k, \theta^i, u_k)\big) \tag{6.47}$$

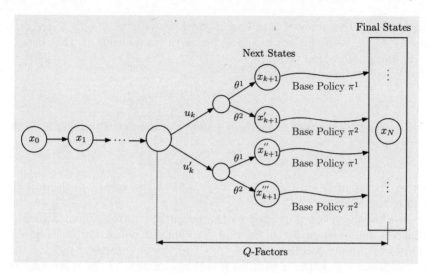

Figure 6.7.1 Schematic illustration of adaptive control by rollout for deterministic systems; cf. Eqs. (6.47) and (6.48). The Q-factors $Q_k(u_k, \theta^i)$ are averaged over θ^i, using the current belief distribution b_k, and the control applied is the one that minimizes the averaged Q-factor $\sum_{i=1}^{m} b_{k,i} Q_k(u_k, \theta^i)$ over $u_k \in U_k(x_k)$.

appearing on the right side of Eq. (6.46) is viewed as a Q-factor that must be computed for every pair (u_k, θ^i), $u_k \in U_k(x_k)$, $i = 1, \ldots, m$, using the base policy π^i. The expected value of this Q-factor,

$$\hat{Q}_k(u_k) = \sum_{i=1}^{m} b_{k,i} Q_k(u_k, \theta^i),$$

must then be calculated for every $u_k \in U_k(x_k)$, and the computation of the rollout control $\tilde{u}_k$ is obtained from the minimization

$$\tilde{u}_k \in \arg \min_{u_k \in U_k(x_k)} \hat{Q}_k(u_k); \tag{6.48}$$

cf. Eq. (6.46). This computation is illustrated in Fig. 6.7.1.

The case of a deterministic system is particularly interesting because we can typically expect that the true parameter θ^* is identified in a finite number of stages, since at each stage k, we are receiving a noiseless measurement relating to θ, namely the state x_k. Once this happens, the problem becomes one of perfect state information.

An illustration similar to the one of Fig. 6.7.1 applies to the rollout scheme (6.44) for the case of a stochastic system. In this case, a Q-factor

$$Q_k(u_k, \theta^i, w_k) = g_k(x_k, \theta^i, u_k, w_k) + \hat{J}^i_{k+1, \pi^i}\big(f_k(x_k, \theta^i, u_k, w_k)\big)$$

must be calculated for every triplet (u_k, θ^i, w_k), using the base policy π^i. The rollout control $\tilde{u}_k$ is obtained by minimizing the expected value of this Q-factor [averaged using the distribution of (θ, w_k)]; cf. Eq. (6.44).

Sec. 6.7   Adaptive Control by Rollout with a POMDP Formulation   179

An interesting and intuitive example that demonstrates the deterministic system case is the popular Worlde puzzle.

Example 6.7.1 (The Wordle Puzzle)

In the classical form of this puzzle, we try to guess a mystery word θ^* out of a known finite collection of 5-letter words. This is done with sequential guesses each of which provides additional information on the correct word θ^*, by using certain given rules to shrink the current mystery list (the smallest list that contains θ^*, based on the currently available information). The objective is to minimize the number of guesses to find θ^* (using more than 6 guesses is considered to be a loss). This type of puzzle descends from the classical family of Mastermind puzzles that centers around decoding a secret sequence of objects (e.g., letters or colors) using partial observations.

The rules for shrinking the mystery list relate to the common letters between the word guesses and the mystery word θ^*, and they will not be described here (there is a large literature regarding the Wordle puzzle). Moreover, θ^* is assumed to be chosen from the initial collection of 5-letter words according to a uniform distribution. Under this assumption, it can be shown that the belief distribution b_k at stage k continues to be uniform over the mystery list. As a result, we may use as state x_k the mystery list at stage k, which evolves deterministically according to an equation of the form (6.45), where u_k is the guess word at stage k. There are several base policies to use in the rollout-like algorithm (6.46), which are described in the paper by Bhambri, Bhattacharjee, and Bertsekas [BBB22], together with computational results that show that the corresponding rollout algorithm (6.46) performs remarkably close to optimal.

The rollout approach also applies to several variations of the Wordle puzzle. Such variations may include for example a larger length $\ell > 5$ of mystery words, and/or a known nonuniform distribution over the initial collection of ℓ-letter words; see [BBB22].

Belief-Based Approximation in Value Space and Rollout

We will now consider an alternative belief-based DP algorithm, given by

$$J'_k(x_k, b_k) = \min_{u_k \in U_k(x_k)} E_{\theta, w_k} \Big\{ g_k(x_k, \theta, u_k, w_k) + J'_{k+1}(x_{k+1}, b_{k+1}) \Big\}, \quad (6.49)$$

where

$$x_{k+1} = f_k(x_k, \theta, u_k, w_k), \qquad b_{k+1} = B_k(x_k, b_k, u_k, x_{k+1}).$$

The relation between the algorithms (6.40) and (6.49) is that if we write b_k as a function of I_k, then the equality

$$J_k(I_k) = J'_k(x_k, b_k)$$

holds as an identity, i.e., for all I_k; see e.g., [Ber17a], Chapter 4.

Let us consider approximations whereby we replace J'_{k+1} with some other function that is more easily computable, in the spirit of approximation in value space. There are several possibilities along this line, some of which have been discussed in previous sections. In particular, in a rollout algorithm we introduce a base policy $\pi = \{\mu_0, \ldots, \mu_{N-1}\}$, with components μ_k that are functions of (x_k, b_k), and we use in place of J'_{k+1}, the cost function $J'_{k+1,\pi}$ of π. This yields the algorithm

$$\tilde{u}_k \in \arg\min_{u_k \in U_k(x_k)} E_{\theta, w_k} \Big\{ g_k(x_k, \theta, u_k, w_k) + J'_{k+1,\pi}\big(x_{k+1}, B_k(x_k, b_k, u_k, x_{k+1})\big) \Big\},$$

where x_{k+1} is given by

$$x_{k+1} = f_k(x_k, \theta, u_k, w_k).$$

Example 6.7.2 (Searching Multiple Sites for a Treasure)

In a classical and challenging problem of search, there are m sites, denoted $1, \ldots, m$, one and only one of which contains a treasure. We assume that searching a site i costs a known amount $c_i > 0$. If site i is searched and a treasure is present at the site, the search reveals it with probability p_i, in which case the process terminates and no further costs are incurred. The problem is to find the search strategy that minimizes the total expected cost to find the treasure.

The basic structure of this problem arises in a broad variety of applications, such as search-and-rescue, inspection-and-repair, as well as various artificial intelligence search contexts. The range of applications is expanded by considering variants of the problem that involve multiple treasures with site-dependent values, and multiple agents/searchers with agent-dependent search costs. Some related problems admit an exact analytical solution. A simple case is discussed in Example 4.3.1 of the DP textbook [Ber17]. A related but structurally different formulation arises in the context of multiarmed bandit problems; see the DP textbook [Ber12], Example 1.3.1.

The probability that site i contains the treasure at the start of stage k, i.e., after k searches, is denoted by $b_{k,i}$. The vector

$$b_k = (b_{k,1}, \ldots, b_{k,m}),$$

is the belief state of the POMDP problem, and the initial belief state b_0 is given. While the treasure has not yet been found, the control u_k at stage k is the choice of the site to search and takes one of the m values $1, \ldots, m$. To be precise, the belief state also includes a state x_k which has two possible values: "treasure not found" and the termination state "treasure found," and there is a trivial system equation whereby x_k transitions to the termination state once a successful search occurs.

The belief states evolves according to an equation that is calculated with the aid of Bayes' rule. In particular, assume that the treasure has not yet been found, and site $\bar{i}$ is searched at time k. Then by applying Bayes' rule, it can be verified that the probability $b_{k,\bar{i}}$ is updated according to

$$b_{k+1,\bar{i}} = \begin{cases} 1 & \text{if the search finds the treasure,} \\ \dfrac{b_{k,\bar{i}}(1-p_{\bar{i}})}{b_{k,\bar{i}}(1-p_{\bar{i}})+\sum_{i\neq \bar{i}} b_{k,i}} & \text{if the search does not find the treasure,} \end{cases}$$

and the probabilities $b_{k,j}$ with $j \neq \bar{i}$ are updated according to

$$b_{k+1,j} = \begin{cases} 0 & \text{if the search finds the treasure,} \\ \dfrac{b_{k,j}}{b_{k,\bar{i}}(1-p_{\bar{i}})+\sum_{i\neq \bar{i}} b_{k,i}} & \text{if the search does not find the treasure.} \end{cases}$$

We write these equations in the abstract form

$$b_{k+1,i} = B_k^i(b_k, u_k, x_{k+1}),$$

where x_{k+1} takes values "treasure found" and "treasure not found" with probabilities $b_{k,\bar{i}}p_{\bar{i}}$ and $1 - b_{k,\bar{i}}p_{\bar{i}}$, respectively ($u_k$ here is the site $\bar{i}$ that is searched at time k).

Suppose now that we have a base policy $\pi = \{\mu_0, \mu_1, \ldots\}$ that consists of functions μ_k, which given the current belief state b_k, choose to search site $\mu_k(b_k)$. Then assuming that the treasure has not yet been found at time k, the rollout algorithm (6.44) takes the form

$$\tilde{u}_k \in \arg\min_{u_k \in \{1,\ldots,m\}} \left[c_{u_k} + E_{x_{k+1}} \left\{ J'_{k+1,\pi}\big(B_k(b_k, u_k, x_{k+1})\big) \right\} \right], \qquad (6.50)$$

where $J'_{k+1,\pi}\big(B_k(b_k, u_k, x_{k+1})\big)$ is the expected cost of the base policy starting from belief state

$$B_k(b_k, u_k, x_{k+1}) = \big(B_k^1(b_k, u_k, x_{k+1}), \ldots, B_k^m(b_k, u_k, x_{k+1})\big).$$

The cost $J'_{k+1,\pi}(b_{k+1})$ appearing in Eq. (6.50), starting from a belief state b_{k+1}, can be computed using the law of iterated expectations as follows: For each $i = 1, \ldots, m$, we compute the cost $C_{i,\pi}$ that would be incurred starting from b_{k+1} and using π, while assuming that the treasure is located at site i; this can be done for example by simulation. We then set

$$J'_{k+1,\pi}(b_{k+1}) = \sum_{i=1}^m b_{k+1,i} C_{i,\pi}.$$

There are several possibilities for base policy π, which are likely context dependent. For a simple example, we may let π be the greedy policy that searches the site $\bar{i}$ of maximum success probability:

$$\bar{i} \in \arg\max_{i \in \{1,\ldots,m\}} b_{k,i}.$$

In conclusion, assuming that the terms

$$E_{x_{k+1}} \left\{ J'_{k+1,\pi}\big(B_k(b_k, u_k, x_{k+1})\big) \right\}$$

are available on-line for all b_k and u_k, through a combination of on-line and off-line computation (that may involve simulation), the algorithm (6.50) admits a tractable implementation, which yields a one-step lookahead suboptimal policy.

6.8 ROLLOUT FOR MINIMAX CONTROL

The problem of optimal control of uncertain systems is usually treated within a stochastic framework, whereby all disturbances $w_0, \ldots, w_{N-1}$ are described by probability distributions, and the expected value of the cost is minimized. However, in many practical situations a stochastic description of the disturbances may not be available, but one may have information with less detailed structure, such as bounds on their magnitude. In other words, one may know a set within which the disturbances are known to lie, but may not know the corresponding probability distribution. Under these circumstances one may use a minimax approach, whereby the worst possible values of the disturbances within the given set are assumed to occur. Within this context, we take the view that the disturbances are chosen by an antagonistic opponent. The minimax approach is also connected with two-player games, when in lack of information about the opponent, we adopt a worst case viewpoint during on-line play, as well as with contexts where we wish to guard against adversarial attacks.†

To be specific, consider a finite horizon context, and assume that the disturbances $w_0, w_1, \ldots, w_{N-1}$ do not have a probabilistic description but rather are known to belong to corresponding given sets $W_k(x_k, u_k) \subset D_k$, $k = 0, 1, \ldots, N-1$, which may depend on the current state x_k and control u_k. The minimax control problem is to find a policy $\pi = \{\mu_0, \ldots, \mu_{N-1}\}$ with $\mu_k(x_k) \in U_k(x_k)$ for all x_k and k, which minimizes the cost function

$$J_\pi(x_0) = \max_{\substack{w_k \in W_k(x_k, \mu_k(x_k)) \\ k=0,1,\ldots,N-1}} \left[g_N(x_N) + \sum_{k=0}^{N-1} g_k(x_k, \mu_k(x_k), w_k) \right].$$

The DP algorithm for this problem takes the following form, which resembles the one corresponding to the stochastic DP problem (maximization is used in place of expectation):

$$J_N^*(x_N) = g_N(x_N), \tag{6.51}$$

$$J_k^*(x_k) = \min_{u_k \in U(x_k)} \max_{w_k \in W_k(x_k, u_k)} \left[g_k(x_k, u_k, w_k) + J_{k+1}^*\big(f_k(x_k, u_k, w_k)\big) \right]. \tag{6.52}$$

This algorithm can be explained by using a principle of optimality type of argument. In particular, we consider the tail subproblem whereby

---

† The minimax approach to decision and control has its origins in the 50s and 60s. It is also referred to by other names, depending on the underlying context, such as *robust control*, *robust optimization*, *control with a set membership description of the uncertainty*, and *games against nature*. In this book, we will be using the minimax control name.

Sec. 6.8 Rollout for Minimax Control

we are at state x_k at time k, and we wish to minimize the "cost-to-go"

$$\max_{\substack{w_t \in W_t(x_t, \mu_t(x_t)) \\ t=k, k+1, \ldots, N-1}} \left[g_N(x_N) + \sum_{t=k}^{N-1} g_t\big(x_t, \mu_t(x_t), w_t\big) \right].$$

We argue that if $\pi^* = \{\mu_0^*, \mu_1^*, \ldots, \mu_{N-1}^*\}$ is an optimal policy for the minimax problem, then the tail of the policy $\{\mu_k^*, \mu_{k+1}^*, \ldots, \mu_{N-1}^*\}$ is optimal for the tail subproblem. The optimal cost of this subproblem is $J_k^*(x_k)$, as given by the DP algorithm (6.51)-(6.52). The algorithm expresses the intuitive fact that when at state x_k at time k, then regardless of what happened in the past, we should choose u_k that minimizes the worst/maximum value over w_k of the sum of the current stage cost plus the optimal cost of the tail subproblem that starts from the next state. This argument requires a mathematical proof, which turns out to involve a few fine points. For a detailed mathematical derivation, we refer to the author's textbook [Ber17a], Section 1.6. However, the DP algorithm (6.51)-(6.52) is correct assuming finite state and control spaces, among other cases.

Approximation in Value Space and Minimax Rollout

The approximation ideas for stochastic optimal control are also relevant within the minimax context. In particular, approximation in value space with one-step lookahead applies at state x_k a control

$$\tilde{u}_k \in \arg\min_{u_k \in U_k(x_k)} \max_{w_k \in W_k(x_k, u_k)} \left[g_k(x_k, u_k, w_k) + \tilde{J}_{k+1}\big(f_k(x_k, u_k, w_k)\big) \right], \quad (6.53)$$

where $\tilde{J}_{k+1}(x_{k+1})$ is an approximation to the optimal cost-to-go $J_{k+1}^*(x_{k+1})$ from state x_{k+1}.

Rollout is obtained when this approximation is the tail cost of some base policy $\pi = \{\mu_0, \ldots, \mu_{N-1}\}$:

$$\tilde{J}_{k+1}(x_{k+1}) = J_{k+1,\pi}(x_{k+1}).$$

Given π, we can compute $J_{k+1,\pi}(x_{k+1})$ by solving a *deterministic maximization* DP problem with the disturbances $w_{k+1}, \ldots, w_{N-1}$ playing the role of "optimization variables/controls." For finite state, control, and disturbance spaces, this is a longest path problem defined on an acyclic graph, since the control variables $u_{k+1}, \ldots, u_{N-1}$ are determined by the base policy. It is then straightforward to implement rollout: at x_k we generate all next states of the form

$$x_{k+1} = f_k(x_k, u_k, w_k)$$

corresponding to all possible values of $u_k \in U_k(x_k)$ and $w_k \in W_k(x_k, u_k)$. We then run the maximization/longest path problem described above to compute $\tilde{J}_{k+1}(x_{k+1})$ from each of these possible next states x_{k+1}. Finally, we obtain the rollout control $\tilde{u}_k$ by solving the minimax problem in Eq. (6.53). Moreover, it is possible to use truncated rollout to approximate the tail cost of the base policy. Note that like all rollout algorithms, the minimax rollout algorithm is well-suited for on-line replanning in problems where data may be changing or may be revealed during the process of control selection.

We noted earlier that deterministic problems allow a more general form of rollout, whereby we may use a base heuristic that need not be a legitimate policy, i.e., it need not be sequentially consistent. For cost improvement it is sufficient that the heuristic be sequentially improving. A similarly more general view of rollout is not easily constructed for stochastic problems, but is possible for minimax control.

In particular, suppose that at any state x_k there is a heuristic that generates a sequence of feasible controls and disturbances, and corresponding states,

$$\{u_k, w_k, x_{k+1}, u_{k+1}, w_{k+1}, x_{k+2}, \ldots, u_{N-1}, w_{N-1}, x_N\},$$

with corresponding cost

$$H_k(x_k) = g_k(x_k, u_k, w_k) + \cdots + g_{N-1}(x_{N-1}, u_{N-1}, w_{N-1}) + g_N(x_N).$$

Then the rollout algorithm applies at state x_k a control

$$\tilde{u}_k \in \arg\min_{u_k \in U_k(x_k)} \max_{w_k \in W_k(x_k, u_k)} \Big[g_k(x_k, u_k, w_k) + H_{k+1}\big(f_k(x_k, u_k, w_k)\big) \Big].$$

This does not preclude the possibility that the disturbances $w_k, \ldots, w_{N-1}$ are chosen by an antagonistic opponent, but allows more general choices of disturbances, obtained for example, by some form of approximate maximization. For example, when the disturbance involves multiple components, $w_k = (w_k^1, \ldots, w_k^m)$, corresponding to multiple opponent agents, *the heuristic may involve an agent-by-agent maximization strategy.*

The sequential improvement condition, similar to the deterministic case, is that for all x_k and k,

$$\min_{u_k \in U_k(x_k)} \max_{w_k \in W_k(x_k, u_k)} \Big[g_k(x_k, u_k, w_k) + H_{k+1}\big(f_k(x_k, u_k, w_k)\big) \Big] \leq H_k(x_k).$$

It guarantees cost improvement, i.e., that for all x_k and k, the rollout policy

$$\tilde{\pi} = \{\tilde{\mu}_0, \ldots, \tilde{\mu}_{N-1}\}$$

satisfies

$$J_{k,\tilde{\pi}}(x_k) \leq H_k(x_k).$$

Thus, generally speaking, minimax rollout is fairly similar to rollout for deterministic as well as stochastic DP problems. The main difference with deterministic (or stochastic) problems is that to compute the Q-factor of a control u_k, we need to solve a maximization problem, rather than carry out a deterministic (or Monte-Carlo, respectively) simulation with the given base policy.

Example 6.8.1 (Pursuit-Evasion Problems)

Consider a pursuit-evasion problem with state $x_k = (x_k^1, x_k^2)$, where x_k^1 is the location of the minimizer/pursuer and x_k^2 is the location of the maximizer/evader, at stage k, in a (finite node) graph defined in two- or three-dimensional space. There is also a cost-free and absorbing termination state that consists of a subset of pairs (x^1, x^2) that includes all pairs with $x^1 = x^2$. The pursuer chooses one out of a finite number of actions $u_k \in U_k(x_k)$ at each stage k, when at state x_k, and if the state is x_k and the pursuer selects u_k, the evader may choose from a known set $X_{k+1}(x_k, u_k)$ of next states x_{k+1}, which depends on (x_k, u_k). The objective of the pursuer is to minimize a nonnegative terminal cost $g(x_N^1, x_N^2)$ at the end of N stages (or reach the termination state, which has cost 0 by assumption). A reasonable base policy for the pursuer can be precomputed by DP as follows: given the current (nontermination) state $x_k = (x_k^1, x_k^2)$, make a move along the path that starts from x_k^1 and minimizes the terminal cost after $N - k$ stages, under the assumption that the evader will stay motionless at his current location x_k^2. (In a variation of this policy, the DP computation is done under the assumption that the evader will follow some nominal sequence of moves.)

For the on-line computation of the rollout control, we need the maximal value of the terminal cost that the evader can achieve starting from every $x_{k+1} \in X_{k+1}(x_k, u_k)$, assuming that the pursuer will follow the base policy (which has already been computed). We denote this maximal value by $\tilde{J}_{k+1}(x_{k+1})$. The required values $\tilde{J}_{k+1}(x_{k+1})$ can be computed by an $(N - k)$-stage DP computation involving the optimal choices of the evader, while assuming the pursuer uses the (already computed) base policy. Then the rollout control for the pursuer is obtained from the minimization

$$\tilde{\mu}_k(x_k) \in \arg \min_{u_k \in U_k(x_k)} \max_{x_{k+1} \in X_{k+1}(x_k, u_k)} \tilde{J}_{k+1}(x_{k+1}).$$

Note that the preceding algorithm can be adapted for the imperfect information case where the pursuer knows x_k^2 imperfectly. This is possible by using a form of assumed certainty equivalence: the pursuer's base policy and the evader's maximization can be computed by using an estimate of the current location x_k^2 instead of the unknown true location.

In the preceding pursuit-evasion example, the choice of the base policy was facilitated by the special structure of the problem. Generally, however, finding a suitable base policy that can be conveniently implemented is an important problem-dependent issue.

Variants of Minimax Rollout

Several of the variants of rollout discussed earlier have analogs in the minimax context, e.g., truncation with terminal cost approximation, multistep and selective step lookahead, and multiagent rollout. In particular, in the ℓ-step lookahead variant, we solve the ℓ-stage problem

$$\min_{u_k,\mu_{k+1},\ldots,\mu_{k+\ell-1}} \max_{\substack{w_t \in W_t(x_t,u_t) \\ t=k,\ldots,k+\ell-1}} \bigg\{ g_k(x_k, u_k, w_k) + \sum_{t=k+1}^{k+\ell-1} g_t\big(x_t, \mu_t(x_t), w_t\big)$$
$$+ H_{k+\ell}(x_{k+\ell}) \bigg\},$$

we find an optimal solution $\tilde{u}_k, \tilde{\mu}_{k+1}, \ldots, \tilde{\mu}_{k+\ell-1}$, and we apply the first component $\tilde{u}_k$ of that solution. As an example, this type of problem is solved at each move of chess programs like AlphaZero, where the terminal cost function is encoded through a position evaluator. In fact when multi-step lookahead is used, special techniques such as *alpha-beta pruning* may be used to accelerate the computations by eliminating unnecessary portions of the lookahead graph. These techniques are well-known in the context of the two-person computer game methodology, and are used widely in games such as chess.

It is interesting to note that, contrary to the case of stochastic optimal control, there is an on-line *constrained form of rollout* for minimax control. Here there are some additional trajectory constraints of the form

$$(x_0, u_0, \ldots, u_{N-1}, x_N) \in C,$$

where C is an arbitrary set. The modification needed is similar to the one of Section 6.6: at partial trajectory

$$\tilde{y}_k = (\tilde{x}_0, \tilde{u}_0, \ldots, \tilde{u}_{k-1}, \tilde{x}_k),$$

generated by rollout, we use a heuristic with cost function H_{k+1} to compute the Q-factor

$$\tilde{Q}_k(\tilde{x}_k, u_k) = \max_{w_k,\ldots,w_{N-1}} \Big[g_k(\tilde{x}_k, u_k, w_k)$$
$$+ H_{k+1}\big(f_k(\tilde{x}_k, u_k, w_k), w_{k+1}, \ldots, w_{N-1}\big) \Big]$$

for each u_k in the set $\tilde{U}_k(\tilde{y}_k)$ that guarantee feasibility [we can check feasibility here by running some algorithm that verifies whether the future disturbances $w_k, \ldots, w_{N-1}$ can be chosen to violate the constraint under the base policy, starting from $(\tilde{y}_k, u_k)$]. Once the set of "feasible controls" $\tilde{U}_k(\tilde{y}_k)$ is computed, we can obtain the rollout control by the Q-factor minimization:

$$\tilde{u}_k \in \arg \min_{u_k \in \tilde{U}_k(\tilde{y}_k)} \tilde{Q}_k(\tilde{x}_k, u_k).$$

We may also use fortified versions of the unconstrained and constrained rollout algorithms, which guarantee a feasible cost-improved rollout policy. This requires the assumption that the base heuristic at the initial state produces a trajectory that is feasible for all possible disturbance sequences. Similar to the deterministic case, there are also truncated and multiagent versions of the minimax rollout algorithm.

Example 6.8.2 (Multiagent Minimax Rollout)

Let us consider a minimax problem where the minimizer's choice involves the collective decision of m agents, $u = (u^1, \ldots, u^m)$, with u^ℓ corresponding to agent ℓ, and constrained to lie within a finite set U^ℓ. Thus u must be chosen from within the set

$$U = U^1 \times \ldots \times U^\ell,$$

which is finite but grows exponentially in size with m. The maximizer's choice w is constrained to belong to a finite set W. We consider multiagent rollout for the minimizer, and for simplicity, we focus on a two-stage problem. However, there are straightforward extensions to a more general multistage framework.

In particular, we assume that the minimizer knowing an initial state x_0, chooses $u = (u^1, \ldots, u^m)$, with $u^\ell \in U^\ell$, $\ell = 1, \ldots, m$, and a state transition

$$x_1 = f_0(x_0, u)$$

occurs with cost $g_0(x_0, u)$. Then the maximizer, knowing x_1, chooses $w \in W$, and a terminal state

$$x_2 = f_1(x_1, w)$$

is generated with cost

$$g_1(x_1, w) + g_2(x_2).$$

The problem is to select $u \in U$, to minimize

$$g_0(x_0, u) + \max_{w \in W} \big[g(x_1, w) + g_2(x_2)\big].$$

The exact DP algorithm for this problem is given by

$$J_1^*(x_1) = \max_{w \in W} \big[g_1(x_1, w) + g_2\big(f_1(x_1, w)\big)\big],$$

$$J_0^*(x_0) = \min_{u \in U} \big[g_0(x_0, u) + J_1^*\big(f_0(x_0, u)\big)\big].$$

This DP algorithm is computationally intractable for large m. The reason is that the set of possible minimizer choices u grows exponentially with m, and for each of these choices the value of $J_1^*\big(f_0(x_0, u)\big)$ must be computed.

However, the problem can be solved approximately with multiagent rollout, using a base policy $\mu = (\mu^1, \ldots, \mu^m)$. Then the number of times $J_1^*\big(f_0(x_0, u)\big)$ needs to be computed is dramatically reduced. This computation is done sequentially, one-agent-at-a-time, as follows:

$$\tilde{u}^1 \in \arg\min_{u^1 \in U^1} \Big[g_0\big(x_0, u^1, \mu^2(x_0), \ldots, \mu^m(x_0)\big)$$
$$+ J_1^*\Big(f_0\big(x_0, u^1, \mu^2(x_0), \ldots, \mu^m(x_0)\big)\Big)\Big],$$

$$\tilde{u}^2 \in \arg\min_{u^2 \in U^2} \Big[g_0\big(x, \tilde{u}^1, u^2, \mu^3(x_0), \ldots, \mu^m(x_0)\big)$$
$$+ J_1^*\Big(f_0\big(x_0, \tilde{u}^1, u^2, \mu^3(x_0), \ldots, \mu^m(x_0)\big)\Big) \Big],$$

$$\cdots \qquad \cdots \qquad \cdots \qquad \cdots$$

$$\tilde{u}^m \in \arg\min_{u^m \in U^m} \Big[g_0\big(x_0, \tilde{u}^1, \tilde{u}^2, \ldots, \tilde{u}^{m-1}, u^m\big)$$
$$+ J_1^*\Big(f_0\big(x_0, \tilde{u}^1, \tilde{u}^2, \ldots, \tilde{u}^{m-1}, u^m\big)\Big) \Big].$$

In this algorithm, the number of times for which $J_1^*\big(f_0(x_0, u)\big)$ must be computed grows linearly with m.

When the number of stages is larger than two, a similar algorithm can be used. Essentially, the one-stage maximizer's cost function J_1^* must be replaced by the optimal cost function of a multistage maximization problem, where the minimizer is constrained to use the base policy.

An interesting question is how do various algorithms work when approximations are used in the min-max and max-min problems? We can certainly improve the minimizer's policy *assuming a fixed policy for the maximizer*. However, it is unclear how to improve both the minimizer's and the maximizer's policies simultaneously. In practice, in *symmetric games*, like chess, a common policy is trained for both players. In particular, in the AlphaZero and TD-Gammon programs this strategy is computationally expedient and has worked well. However, there is no reliable theory to guide the simultaneous training of policies for both maximizer and minimizer, and it is quite plausible that unusual behavior may arise in exceptional cases. Even *exact* policy iteration methods for Markov games encounter serious convergence difficulties, and need to be modified for reliable behavior. The author's paper [Ber21c] and book [Ber22a] (Chapter 5) address these convergence issues with modified versions of the policy iteration method, and give many earlier references.

We finally note another source of difficulty in minimax control: Newton's method applied to solution of the Bellman equation for minimax problems exhibits more complex behavior than its expected value counterpart. The reason is that the Bellman operator T for infinite horizon problems, given by

$$(TJ)(x) = \min_{u \in U(x)} \max_{w \in W(x,u)} \Big[g(x, u, w) + \alpha J\big(f(x, u, w)\big) \Big], \qquad \text{for all } x,$$

is neither convex nor concave as a function of J. To see this, note that the function

$$\max_{w \in W(x,u)} \Big[g(x, u, w) + \alpha J\big(f(x, u, w)\big) \Big],$$

viewed as a function of J [for fixed (x, u)], is convex, and when minimized over $u \in U(x)$, it becomes neither convex nor concave (cf. Fig. 3.9.4).

Sec. 6.8 Rollout for Minimax Control

As a result there are special difficulties in connection with convergence of Newton's method and the natural form of policy iteration, given by Pollatschek and Avi-Itzhak [PoA69]; see also Chapter 5 of the author's abstract DP book [Ber22a].

Minimax Control and Zero-Sum Game Theory

Zero-sum game problems are viewed as fundamental in the field of economics, and there is an extensive and time-honored theory around them. In the case where the game involves a dynamic system

$$x_{k+1} = f_k(x_k, u_k, w_k),$$

and a cost function

$$g_k(x_k, u_k, w_k),$$

there are two players, the minimizer choosing $u_k \in U_k(x_k)$, and the maximizer choosing $w_k \in W_k(x_k)$, at each stage k. Such zero-sum games involve *two* minimax control problems:

(a) The *min-max problem*, where the minimizer chooses a policy first and the maximizer chooses a policy second with knowledge of the minimizer's policy. The DP algorithm for this problem has the form

$$J_N^*(x_N) = g_N(x_N),$$

$$J_k^*(x_k) = \min_{u_k \in U_k(x_k)} \max_{w_k \in W_k(x_k)} \Big[g_k(x_k, u_k, w_k) + J_{k+1}^*\big(f_k(x_k, u_k, w_k)\big) \Big].$$

(b) The *max-min problem*, where the maximizer chooses policy first and the minimizer chooses policy second with knowledge of the maximizer's policy. The DP algorithm for this problem has the form

$$\hat{J}_N(x_N) = g_N(x_N),$$

$$\hat{J}_k(x_k) = \max_{w_k \in W_k(x_k)} \min_{u_k \in U_k(x_k)} \Big[g_k(x_k, u_k, w_k) + \hat{J}_{k+1}\big(f_k(x_k, u_k, w_k)\big) \Big].$$

A basic and easily seen fact is that

Max-Min optimal value $\leq$ Min-Max optimal value.

Game theory is particularly interested on conditions that guarantee that

Max-Min optimal value $=$ Min-Max optimal value. (6.54)

However, this question is of limited interest in engineering contexts that involve worst case design. Moreover, the validity of the minimax equality (6.54) is beyond the range of practical RL. This is so primarily because once approximations are introduced, the delicate assumptions that guarantee this equality are disrupted.

6.9 SMALL STAGE COSTS AND LONG HORIZON - CONTINUOUS-TIME ROLLOUT

Let us consider the deterministic one-step approximation in value space scheme

$$\tilde{\mu}_k(x_k) \in \arg\min_{u_k \in U_k(x_k)} \left[g_k(x_k, u_k) + \tilde{J}_{k+1}\big(f_k(x_k, u_k)\big) \right]. \tag{6.55}$$

In the context of rollout, $\tilde{J}_{k+1}\big(f_k(x_k, u_k)\big)$ is either the cost of the trajectory generated by the base heuristic starting from the next state $f_k(x_k, u_k)$, or some approximation that may involve truncation and terminal cost function approximation, as in the truncated rollout scheme of Section 6.5.

There is a special difficulty within this context, which is often encountered in practice. It arises when the cost per stage $g_k(x_k, u_k)$ is either 0 or is small relative to the cost-to-go approximation $\tilde{J}_{k+1}\big(f_k(x_k, u_k)\big)$. Then there is a potential pitfall to contend with: *the cost approximation errors that are inherent in the term $\tilde{J}_{k+1}\big(f_k(x_k, u_k)\big)$ may overwhelm the first stage cost term $g_k(x_k, u_k)$*, with unpredictable consequences for the quality of the one-step-lookahead policy $\tilde{\pi} = \{\tilde{\mu}_0, \ldots, \tilde{\mu}_{N-1}\}$. We will discuss this difficulty by first considering a discrete-time problem arising from discretization of a continuous-time optimal control problem.

Continuous-Time Optimal Control and Approximation in Value Space

Consider a problem that involves a vector differential equation of the form

$$\dot{x}(t) = h\big(x(t), u(t), t\big), \qquad 0 \le t \le T, \tag{6.56}$$

where $x(t) \in \Re^n$ is the state vector at time t, $\dot{x}(t) \in \Re^n$ is the vector of first order time derivatives of the state at time t, $u(t) \in U \subset \Re^m$ is the control vector at time t, where U is the control constraint set, and T is a given terminal time. Starting from a given initial state $x(0)$, we want to find a feasible control trajectory $\{u(t) \mid t \in [0, T]\}$, which together with its corresponding state trajectory $\{x(t) \mid t \in [0, T]\}$, minimizes a cost function of the form

$$G\big(x(T)\big) + \int_0^T g\big(x(t), u(t), t\big) dt, \tag{6.57}$$

where g represents cost per unit time, and G is a terminal cost function. This is a classical problem with a long history.

Let us consider a simple conversion of the preceding continuous-time problem to a discrete-time problem, while treading lightly over some of the associated mathematical fine points. We introduce a small discretization increment $\delta > 0$, such that $T = \delta N$ where N is a large integer, and we replace the differential equation (6.56) by

$$x_{k+1} = x_k + \delta \cdot h_k(x_k, u_k), \qquad k = 0, \ldots, N-1.$$

Sec. 6.9　Small Stage Costs and Long Horizon -Continuous-Time Rollout

Here the function h_k is given by

$$h_k(x_k, u_k) = h\big(x(k\delta), u(k\delta), k\delta\big),$$

where we view $\{x_k \mid k = 0, \ldots, N-1\}$ and $\{u_k \mid k = 0, \ldots, N-1\}$ as state and control trajectories, respectively, which approximate the corresponding continuous-time trajectories:

$$x_k \approx x(k\delta), \qquad u_k \approx u(k\delta).$$

We also replace the cost function (6.57) by

$$g_N(x_N) + \sum_{k=0}^{N-1} \delta \cdot g_k(x_k, u_k),$$

where

$$g_N(x_N) = G\big(x(N\delta)\big), \qquad g_k(x_k, u_k) = g\big(x(k\delta), u(k\delta), k\delta\big).$$

Thus the approximation in value space scheme with time discretization takes the form

$$\tilde{\mu}_k(x_k) \in \arg\min_{u_k \in U} \Big[\delta \cdot g_k(x_k, u_k) + \tilde{J}_{k+1}\big(x_k + \delta \cdot h_k(x_k, u_k)\big)\Big]; \quad (6.58)$$

where $\tilde{J}_{k+1}$ is the function that approximates the cost-to-go starting from a state at time $k+1$. We note here that the ratio of the terms $\delta \cdot g_k(x_k, u_k)$ and $\tilde{J}_{k+1}\big(x_k + \delta \cdot h_k(x_k, u_k)\big)$ is likely to tend to 0 as $\delta \to 0$, since $\tilde{J}_{k+1}\big(x_k + \delta \cdot h_k(x_k, u_k)\big)$ ordinarily stays roughly constant at a nonzero level as $\delta \to 0$. This suggests that the one-step lookahead minimization may be degraded substantially by discretization, and other errors, including rollout truncation and terminal cost approximation. Note that a similar sensitivity to errors may occur in other discrete-time models that involve frequent selection of decisions, with cost per stage that is very small relative to the cumulative cost over many stages and/or the terminal cost.

To deal with this difficulty, we subtract the constant $\tilde{J}_k(x_k)$ in the one-step-lookahead minimization (6.58), and write

$$\tilde{\mu}_k(x_k) \in \arg\min_{u_k \in U} \Big[\delta \cdot g_k(x_k, u_k) + \big(\tilde{J}_{k+1}\big(x_k + \delta \cdot h_k(x_k, u_k)\big) - \tilde{J}_k(x_k)\big)\Big]; \quad (6.59)$$

since $\tilde{J}_k(x_k)$ does not depend on u_k, the results of the minimization are not affected. Assuming $\tilde{J}_k$ is differentiable with respect to its argument, we can write

$$\tilde{J}_{k+1}\big(x_k + \delta \cdot h_k(x_k, u_k)\big) - \tilde{J}_k(x_k) \approx \delta \cdot \nabla_x \tilde{J}_k(x_k)' h_k(x_k, u_k),$$

where $\nabla_x \tilde{J}_k$ denotes the gradient of J_k (a column vector), and prime denotes transposition. By dividing with δ, and taking informally the limit as $\delta \to 0$, we can write the one-step lookahead minimization (6.59) as

$$\tilde{\mu}(t) \in \arg\min_{u(t) \in U} \Big[g\big(x(t), u(t), t\big) + \nabla_x \tilde{J}_t\big(x(t)\big)' h\big(x(t), u(t), t\big) \Big], \quad (6.60)$$

where $\tilde{J}_t(x)$ is the continuous-time cost function approximation and $\nabla_x \tilde{J}_t(x)$ is its gradient with respect to x. This is the correct analog of the approximation in value space scheme (6.55) for continuous-time problems.

Rollout for Continuous-Time Optimal Control

In view of the value approximation scheme of Eq. (6.60), it is natural to speculate that the continuous-time analog of rollout with a base policy of the form

$$\pi = \Big\{ \mu_t\big(x(t)\big) \mid 0 \le t \le T \Big\}, \quad (6.61)$$

where $\mu_t\big(x(t)\big) \in U$ for all $x(t)$ and t, has the form

$$\tilde{\mu}_t\big(x(t)\big) \in \arg\min_{u(t) \in U} \Big[g\big(x(t), u(t), t\big) + \nabla_x J_{\pi,t}\big(x(t)\big)' h\big(x(t), u(t), t\big) \Big]. \quad (6.62)$$

Here $J_{\pi,t}\big(x(t)\big)$ is the cost of the base policy π starting from state $x(t)$ at time t, and satisfies the terminal condition

$$J_{\pi,T}\big(x(T)\big) = G\big(x(T)\big).$$

Computationally, the inner product in the right-hand side of the above minimization can be approximated using the finite difference formula

$$\nabla_x J_{\pi,t}\big(x(t)\big)' h\big(x(t), u(t), t\big) \approx \frac{J_{\pi,t}\Big(x(t) + \delta \cdot h\big(x(t), u(t), t\big)\Big) - J_{\pi,t}\big(x(t)\big)}{\delta},$$

which can be calculated by running the base policy π starting from $x(t)$ and from $x(t) + \delta \cdot h\big(x(t), u(t), t\big)$. (This finite differencing operation may involve tricky computational issues, but we will not get into this.)

An important question is how to select the base policy π. A choice that is often sensible and convenient is to choose π to be a "short-sighted" policy, which takes into account the "short term" cost from the current state (say for a very small horizon starting from the current time t), but ignores the remaining cost. An extreme case is the *myopic* policy, given by

$$\mu_t\big(x(t)\big) \in \arg\min_{u \in U} g\big(x(t), u(t), t\big).$$

Sec. 6.9 Small Stage Costs and Long Horizon -Continuous-Time Rollout

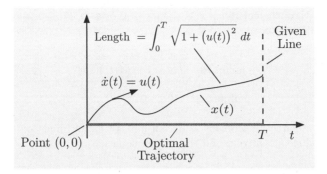

Figure 6.9.1 Problem of finding a curve of minimum length from a given point to a given line, and its formulation as a calculus of variations problem.

This policy is the continuous-time analog of the greedy policy that we discussed in the context of discrete-time problems, and the traveling salesman Example 6.4.1 in particular.

The following example illustrates the rollout algorithm (6.62) with a problem where the base policy cost $J_{\pi,t}(x(t))$ is independent of $x(t)$ (it depends only on t), so that

$$\nabla_x J_{\pi,t}(x(t)) \equiv 0.$$

In this case, in view of Eq. (6.60), the rollout policy is myopic. It turns out that the optimal policy in this example is also myopic, so that the rollout policy is optimal, even though the base policy is very poor.

Example 6.9.1 (A Calculus of Variations Problem)

This is a simple example from the classical context of calculus of variations (see [Ber17a], Example 7.1.3). The problem is to find a minimum length curve that starts at a given point and ends at a given line. Without loss of generality, let $(0,0)$ be the given point, and let the given line be the vertical line that passes through $(T,0)$, as shown in Fig. 6.9.1.

Let $(t, x(t))$ be the points of the curve, where $0 \leq t \leq T$. The portion of the curve joining the points $(t, x(t))$ and $(t+dt, x(t+dt))$ can be approximated, for small dt, by the hypotenuse of a right triangle with sides dt and $\dot{x}(t)dt$. Thus the length of this portion is

$$\sqrt{(dt)^2 + (\dot{x}(t))^2 (dt)^2},$$

which is equal to

$$\sqrt{1 + (\dot{x}(t))^2}\, dt.$$

The length of the entire curve is the integral over $[0, T]$ of this expression, so the problem is to

$$\text{minimize} \int_0^T \sqrt{1 + \big(\dot{x}(t)\big)^2}\, dt$$

$$\text{subject to}\ x(0) = 0.$$

To reformulate the problem as a continuous-time optimal control problem, we introduce a control u and the system equation

$$\dot{x}(t) = u(t), \qquad x(0) = 0.$$

Our problem then takes the form

$$\text{minimize} \int_0^T \sqrt{1 + \big(u(t)\big)^2}\, dt.$$

This is a problem that fits our continuous-time optimal control framework, with

$$h\big(x(t), u(t), t\big) = u(t), \quad g\big(x(t), u(t), t\big) = \sqrt{1 + \big(u(t)\big)^2}, \quad G\big(x(T)\big) = 0.$$

Consider now a base policy π whereby the control depends only on t and not on x. Such a policy has the form

$$\mu_t\big(x(t)\big) = \beta(t), \qquad \text{for all } x(t),$$

where $\beta(t)$ is some scalar function. For example, $\beta(t)$ may be constant, $\beta(t) \equiv \bar{\beta}$ for some scalar $\bar{\beta}$, which yields a straight line trajectory that starts at $(0,0)$ and makes an angle ϕ with the horizontal with $\tan(\phi) = \bar{\beta}$. The cost function of the base policy is

$$J_{\pi,t}\big(x(t)\big) = \int_t^T \sqrt{1 + \beta(\tau)^2}\, d\tau,$$

which is independent of $x(t)$, so that $\nabla_x J_{\pi,t}\big(x(t)\big) \equiv 0$. Thus, from the minimization of Eq. (6.62), we have

$$\tilde{\mu}_t\big(x(t)\big) \in \arg\min_{u(t)\in\Re} \sqrt{1 + \big(u(t)\big)^2},$$

and the rollout policy is

$$\tilde{\mu}_t\big(x(t)\big) \equiv 0.$$

This is the optimal policy: it corresponds to the horizontal straight line that starts at $(0,0)$ and ends at $(T,0)$.

Rollout with General Base Heuristics - Sequential Improvement

An extension of the rollout algorithm (6.62) is to use a more general base heuristic whose cost function $H_t(x(t))$ can be evaluated by simulation. This rollout algorithm has the form

$$\tilde{\mu}(t) \in \arg\min_{u(t) \in U}\left[g(x(t), u(t), t) + \nabla_x H_t(x(t))' h(x(t), u(t), t)\right].$$

Here the policy cost function $J_{\pi,t}$ is replaced by a more general differentiable function H_t, obtainable through a base heuristic, which may lack the sequential consistency property that is inherent in policies.

We will now show a cost improvement property of the rollout algorithm based on the natural condition

$$H_T(\tilde{x}(T)) = G(\tilde{x}(T)), \tag{6.63}$$

and the assumption

$$\min_{u(t) \in U}\left[g(x(t), u(t), t) + \nabla_t H_t(x(t)) + \nabla_x H_t(x(t))' h(x(t), u(t), t)\right] \leq 0, \tag{6.64}$$

for all $(x(t), t)$, where $\nabla_x H_t$ denotes gradient with respect to x, and $\nabla_t H_t$ denotes gradient with respect to t. This assumption is the continuous-time analog of the sequential improvement condition of Definition 6.4.2 [cf. Eq. (6.18)]. Under this assumption, we will show that

$$J_{\tilde{\pi},0}(x(0)) \leq H_0(x(0)), \tag{6.65}$$

i.e., the cost of the rollout policy starting from the initial state $x(0)$ is no worse than the base heuristic cost starting from the same initial state.

Indeed, let $\{\tilde{x}(t) \mid t \in [0,T]\}$ and $\{\tilde{u}(t) \mid t \in [0,T]\}$ be the state and control trajectories generated by the rollout policy starting from $x(0)$. Then the sequential improvement condition (6.64) yields

$$g(\tilde{x}(t), \tilde{u}(t), t) + \nabla_t H_t(\tilde{x}(t)) + \nabla_x H_t(\tilde{x}(t))' h(\tilde{x}(t), \tilde{u}(t), t) \leq 0$$

for all t, and by integration over $[0, T]$, we obtain

$$\int_0^T g(\tilde{x}(t), \tilde{u}(t), t)\, dt + \int_0^T \left(\nabla_t H_t(\tilde{x}(t)) + \nabla_x H_t(\tilde{x}(t))' h(\tilde{x}(t), \tilde{u}(t), t)\right) dt \leq 0. \tag{6.66}$$

The second integral above can be written as

$$\int_0^T \left(\nabla_t H_t(\tilde{x}(t)) + \nabla_x H_t(\tilde{x}(t))' h(\tilde{x}(t), \tilde{u}(t), t)\right) dt$$

$$= \int_0^T \left(\nabla_t H_t(\tilde{x}(t)) + \nabla_x H_t(\tilde{x}(t))' \frac{d\tilde{x}(t)}{dt}\right) dt,$$

and its integrand is the total differential with respect to time: $\frac{d}{dt}\big(H_t(\tilde{x}(t))\big)$. Thus we obtain from Eq. (6.66)

$$\int_0^T g\big(\tilde{x}(t),\tilde{u}(t),t\big)\,dt + \int_0^T \frac{d}{dt}\big(H_t(\tilde{x}(t))\big)\,dt$$
$$= \int_0^T g\big(\tilde{x}(t),\tilde{u}(t),t\big)\,dt + H_T\big(\tilde{x}(T)\big) - H_0\big(\tilde{x}(0)\big) \leq 0. \qquad (6.67)$$

Since $H_T\big(\tilde{x}(T)\big) = G\big(\tilde{x}(T)\big)$ [cf. Eq. (6.63)] and $\tilde{x}(0) = x(0)$, from Eq. (6.67) [which is a direct consequence of the sequential improvement condition (6.64)], it follows that

$$J_{\tilde{\pi},0}\big(x(0)\big) = \int_0^T g\big(\tilde{x}(t),\tilde{u}(t),t\big)\,dt + G\big(\tilde{x}(T)\big) \leq H_0\big(x(0)\big),$$

thus proving the cost improvement property (6.65).

Note that the sequential improvement condition (6.64) is satisfied if H_t is the cost function $J_{\pi,t}$ corresponding to a base policy π. The reason is that for any policy $\pi = \{\mu_t(x(t)) \mid 0 \leq t \leq T\}$ [cf. Eq. (6.61)], the analog of the DP algorithm (under the requisite mathematical conditions) takes the form

$$0 = g\big(x(t),\mu_t(x(t)),t\big) + \nabla_t J_{\pi,t}\big(x(t)\big) + \nabla_x J_{\pi,t}\big(x(t)\big)' h\big(x(t),\mu_t(x(t)),t\big). \qquad (6.68)$$

In continuous-time optimal control theory, this is known as the *Hamilton-Jacobi-Bellman equation*. It is a partial differential equation, which may be viewed as the continuous-time analog of the DP algorithm for a single policy; there is also a Hamilton-Jacobi-Bellman equation for the optimal cost function $J_t^*\big(x(t)\big)$ (see optimal control textbook accounts, such as [Ber17a], Section 7.2, and the references cited there). As illustration, the reader may verify that the cost function of the base policy used in the calculus of variations problem of Example 6.9.1 satisfies this equation. It can be seen from the Hamilton-Jacobi-Bellman Eq. (6.68) that when $H_t = J_{\pi,t}$, the sequential improvement condition (6.64) and the cost improvement property (6.65) hold.

Approximating Cost Function Differences

Let us finally note that the preceding analysis suggests that when dealing with a discrete-time problem with a long horizon N, a system equation $x_{k+1} = f_k(x_k, u_k)$, and a small cost per stage $g_k(x_k, u_k)$ relative to the optimal cost-to-go function $J_{k+1}\big(f_k(x_k, u_k)\big)$, it is worth considering an alternative implementation of the approximation in value space scheme. In particular, we should consider approximating the cost differences

$$D_k(x_k, u_k) = J_{k+1}\big(f_k(x_k, u_k)\big) - J_k(x_k)$$

instead of approximating the cost-to-go functions $J_{k+1}\bigl(f_k(x_k, u_k)\bigr)$. The one-step-lookahead minimization (6.55) should then be replaced by

$$\tilde{\mu}_k(x_k) \in \arg \min_{u_k \in U_k(x_k)} \bigl[g_k(x_k, u_k) + \tilde{D}_k(x_k, u_k)\bigr],$$

where $\tilde{D}_k$ is the approximation to D_k.

Note also that while for continuous-time problems, the idea of approximating the gradient of the optimal cost function is essential and comes out naturally from the analysis, for discrete-time problems, approximating cost-to-go differences rather than cost functions is optional and should be considered in the context of a given problem. Methods along this line include advantage updating, cost shaping, biased aggregation, and the use of baselines, for which we refer to the books [BeT96], [Ber19a], and [Ber20a]. A special method to explicitly approximate cost function differences is *differential training*, which is discussed in Section 4.3.4 of the book [Ber20a].

Unfortunately, approximating cost-to-go differences may not be effective when the cost per stage is 0 for all states, while a nonzero cost is incurred only at termination. This type of cost structure occurs, among others, in games such as chess and backgammon. In this case a potentially effective remedy is to resort to longer lookahead, either through multistep lookahead minimization, or through some form of truncated rollout, as it is done in the AlphaZero and TD-Gammon programs.

6.10 EPILOGUE

While the ideas of approximation in value space, rollout, and PI have a long history, their significance has been highlighted by the success of AlphaZero, and the earlier but just as impressive TD-Gammon program. Both programs were trained off-line extensively using sophisticated approximate PI algorithms and neural networks. Yet in AlphaZero, the player obtained off-line was greatly improved by on-line play, as we have discussed in Chapter 1. Moreover, TD-Gammon was greatly improved by supplementing its on-line play scheme with truncated rollout.

We have argued that this performance enhancement by on-line play defines a new transformative and broadly applicable paradigm for decision and control, which is couched on the AlphaZero/TD-Gammon design principles: on-line decision making, using approximation in value space with multistep lookahead, and rollout. Moreover, this paradigm provides the basis for a much needed unification of the methodological cultures of reinforcement learning, and optimization/control, and particularly MPC, which in fact embodies several of the AlphaZero/TD-Gammon design ideas.

We have highlighted the multiple beneficial properties of truncated rollout as a reliable, easily implementable, and cost effective alternative to long lookahead minimization. We have also noted how rollout with a stable policy, enhances the stability properties of the controller obtained by approximation in value space schemes. The issue of stability is of paramount importance in control system design and MPC, but is not addressed adequately by the RL methodology, as practiced by the AI community.

Moreover, we have argued that there is an additional benefit of policy improvement by approximation in value space, not observed in the context of games (which have unchanging rules and environment). It is well-suited for on-line replanning and changing problem parameters, as in the context of indirect adaptive control.

The Mathematical Framework

From a mathematical point of view, we have aimed to provide the framework and insights, which facilitate the use of on-line decision making on top of off-line training. In particular, through a unified abstract DP analysis, which is well-suited to visualization, we have shown that the principal ideas of approximation in value space and rollout apply very broadly to deterministic and stochastic optimal control problems, involving both discrete and continuous search spaces.

A key idea of this work is *the interpretation of approximation in value space with one-step lookahead as a step of Newton's method*. This idea has been known for a long time, but only within the more restrictive context of policy iteration, where the cost function approximation is restricted to be the cost function of some policy. The extensions of this idea, including more general cost function approximations, multistep lookahead, truncated rollout, connection with stability issues, and discrete and multiagent optimization, which are provided in this work, are new (following their introduction in the book [Ber20a]), and aim to promote the view that Newton's method and other classical algorithms, such as Newton-SOR, are central conceptual elements of the RL methodology.

Mathematical proofs of our superlinear convergence rate and sensitivity results were given primarily for the case of a one-dimensional quadratic problem (Chapter 4). However, these results can be straightforwardly extended to more general multidimensional linear quadratic problems. Moreover, similar results can be obtained for more general problems, by using the equivalence to a Newton step that we showed in Chapter 3, and by relying on known analyses of nondifferentiable forms of Newton's method, which we have discussed in the Appendix and in the paper [Ber22b]. At the same time considerable work remains to be done to clarify exceptional behaviors of Newton's method within various DP contexts, and to work out rigorously the associated mathematical results. Furthermore, there is a need for better characterization of the region of attraction of the method in contexts beyond the nice discounted problems with bounded cost per stage.

Sec. 6.10   Epilogue   199

A major supplementary idea of this work is *the interpretation of off-line training of policies and cost approximations as means for enhancement of the initial condition of the Newton step*. Among others, this interpretation supports the view that the Newton step/on-line player is the key determinant of the overall scheme's performance, and that the initial condition adjustment/off-line training plays a subsidiary role. Still, however, while this is a valuable conceptual starting point, we expect that in specific contexts the initial condition of the Newton step and the attendant off-line training process may play an important role. For example in the MPC context, off-line training may be critical in dealing with issues of stability, state constraints, and target tube construction.

We finally note that the preceding mathematical ideas have a universal character in view of their abstract DP foundation, which allows their use in very general DP contexts, involving both discrete and continuous state and control spaces, as well as value and policy approximations. As a result, they can be effectively integrated within a broad range of methodologies, such as adaptive control, MPC, decentralized and multiagent control, discrete and Bayesian optimization, neural network-based approximations, and heuristic algorithms for discrete optimization, as we have discussed in greater detail in the books [Ber19a] and [Ber20a].

Rollout for Finite-Horizon Problems and Discrete Optimization

In the present chapter we have noted that while our starting point in this work has been infinite horizon problems, approximation in value space and rollout can be applied to finite horizon problems as well, and can be similarly interpreted in terms of Newton's method. In particular, finite horizon problems can be converted to infinite horizon stochastic shortest path problems with a termination state that corresponds to the end of the horizon. Once this is done, the conceptual framework of our work can be applied to provide insight on the connections between approximation in value space, rollout, and Newton's method.

Thus, our rollout ideas find application beyond the infinite horizon DP context, and apply to the solution of classical discrete and combinatorial optimization problems, as we have aimed to show in this chapter. This was the basis for the original proposal of the use of rollout for discrete and combinatorial optimization problems in the paper by Bertsekas, Tsitsiklis, and Wu [BTW97]; see also the author's book [Ber20a], which provides a fuller presentation of the finite horizon methodology.

The book [Ber20a] also contains several examples of application of rollout to discrete optimization and provides references to many works spanning the period from the late 90s to the present. These works discuss variants and problem-specific adaptations of rollout algorithms for a broad variety of practical problems, and consistently report favorable computational experience. The size of the cost improvement over the base policy is often impressive, evidently owing to the fast convergence rate of Newton's method that underlies rollout. Moreover these works illustrate some of the other important advantages of rollout: reliability, simplicity, suitability for on-line replanning, and the ability to interface with other RL techniques, such as neural network training, which can be used to provide suitable base policies and/or approximations to their cost functions.

We finally note that in Section 6.5.2 we have explored yet another use of rollout, as a means for facilitating approximation in value space with long multistep lookahead, by performing expeditiously (yet approximately) the multistep lookahead minimization over the corresponding acyclic lookahead graph. This idea is new and the range of applications of the incremental rollout algorithm as a competitor to Monte Carlo Tree Search and other tree search methods, is very broad.

APPENDIX A:
Newton's Method and Error Bounds

Contents
A.1. Newton's Method for Differentiable Fixed Point Problems . . . . p. 202
A.2. Newton's Method Without Differentiability of the Bellman Operator . . . . p. 207
A.3. Local and Global Error Bounds for Approximation in Value Space . . . . p. 210
A.4. Local and Global Error Bounds for Approximate Policy Iteration . . . . p. 212

In this appendix, we first develop the classical theory of Newton's method for solving a fixed point problem of the form

$$y = G(y),$$

where y is an n-dimensional vector, and $G : \Re^n \mapsto \Re^n$ is a continuously differentiable mapping.† We then extend the results to the case where G is nondifferentiable because it is obtained as the minimum of continuously differentiable mappings, as in the case of the Bellman operator (cf. Chapter 3).

The convergence analysis relates to the solution of Bellman's equation $J = TJ$, for the case where J is an n-dimensional (there are n states), TJ is real-valued for all real-valued J, and T is either differentiable or involves minimization over a finite number of controls. However, the analysis illuminates the mechanism by which Newton's method works for more general problems, involving for example infinite spaces problems. Moreover, the analysis does not use the concavity and monotonicity properties of T that hold in the DP contexts that we have discussed (discounted, SSP, and nonnegative-cost deterministic problems, cf. Chapter 2), but may not hold in other DP-related contexts.

A.1 NEWTON'S METHOD FOR DIFFERENTIABLE FIXED POINT PROBLEMS

Newton's method is an iterative algorithm that generates a sequence $\{y_k\}$, starting from some initial vector y_0. It aims to obtain asymptotically a fixed point of G, i.e., $y_k \to y^*$, where y^* is such that $y^* = G(y^*)$. Newton's method is usually analyzed in the context of solving systems of equations. In particular, by introducing the mapping $H : \Re^n \mapsto \Re^n$, given by

$$H(y) = G(y) - y, \qquad y \in \Re^n,$$

the fixed point problem is transformed to solving the equation $H(y) = 0$. We view $H(y)$ as a column vector in $\Re^n$, with its n components denoted by $H_1(y), \ldots, H_n(y)$:

$$H(y) = \begin{pmatrix} H_1(y) \\ \vdots \\ H_n(y) \end{pmatrix}.$$

Each of the functions H_i is given by

$$H_i(y) = G_i(y) - y_i,$$

---

† The subsequent analysis of Sections A.1 and A.2 also holds when G maps an open subset Y of $\Re^n$ into itself, i.e., $G(y) \in Y$ for all $y \in Y$.

Sec. A.1   Newton's Method for Differentiable Fixed Point Problems   203

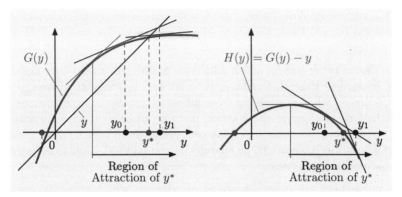

Figure A.1 Illustration of Newton's method for solving the differentiable fixed point problem $y = G(y)$, and equivalently, the equation $H(y) = 0$ where

$$H(y) = G(y) - y.$$

At each iteration the method first linearizes the problem at the current iterate y_k via a first order Taylor series expansion, and then computes y_{k+1} as the solution of the linearized problem. The method converges to a solution y^* when started within its region of attraction, i.e, the set of starting points y_0 such that

$$\|y_{k+1} - y^*\| \leq \|y_k - y^*\|, \qquad \text{and } y_k \to y^*.$$

In this one-dimensional case where G is concave and monotonically increasing, the region of attraction is as shown.

where y_i is the ith component of y and G_i is the ith component of G.

Assuming that H is differentiable, Newton's method takes the form

$$y_{k+1} = y_k - \left(\nabla H(y_k)'\right)^{-1} H(y_k), \tag{A.1}$$

where ∇H is the $n \times n$ matrix whose columns are the gradients $\nabla H_1, \ldots, \nabla H_n$ of the n components $H_1, \ldots, H_n$, viewed as column vectors:

$$\nabla H(y) = \left(\nabla H_1(y) \ \cdots \ \nabla H_n(y)\right),$$

and $\nabla H(y_k)'$ denotes the transpose of $\nabla H(y_k)$ [i.e., $\nabla H(y_k)'$ is the Jacobian of H at y_k]. The algorithm (A.1), illustrated in one dimension in Fig. A.1, is the *classical form of Newton's method*, and it assumes that $\nabla H(y_k)$ is invertible for every k.†

Its analysis has two principal aspects:

---

† An intuitive view of the Newton iteration is that it first linearizes H at the current point y_k via a first order Taylor series expansion,

$$H(y) \approx H(y_k) + \nabla H(y_k)'(y - y_k),$$

(a) *Local convergence*, which deals with the behavior near a nonsingular solution y^*, i.e., one where $H(y^*) = 0$ and the matrix $\nabla H(y^*)$ is invertible.

(b) *Global convergence*, which addresses modifications that are necessary to ensure that the method is valid and is likely to converge to a solution when started far from all solutions. Such modifications may include changes in the starting point y_0 to bring it within the region of convergence (the set of starting points from which convergence to a solution is assured), or changes in the method itself to improve its stability properties.

Of course there are other questions of interest, which relate to the convergence and rate of convergence aspects of the method. The literature of the subject is very extensive, and is covered in several books and research papers.

In this appendix, we will only consider local convergence questions, and we will focus on a single nonsingular solution, i.e., a vector y^* such that $H(y^*) = 0$ and $\nabla H(y^*)$ is invertible. The principal result here is that the Newton method (A.1) converges superlinearly when started close enough to y^*. For a simple argument that shows this fact, suppose that the method generates a sequence $\{y_k\}$ that converges to y^*. Let us use a first order expansion around y_k to write

$$0 = H(y^*) = H(y_k) + \nabla H(y_k)'(y^* - y_k) + o(\|y_k - y^*\|).$$

and then computes y_{k+1} as the solution of the *linearized system*

$$H(y_k) + \nabla H(y_k)'(y - y_k) = 0.$$

Equivalently, in terms of the original fixed point problem $y = G(y)$, Newton's method first linearizes G at the current point y_k via a first order Taylor series expansion,

$$G(y) \approx G(y_k) + \nabla G(y_k)'(y - y_k),$$

where $\nabla G(y_k)'$ is the Jacobian matrix of G evaluated at y_k. It then computes y_{k+1} as the solution of the linearized fixed point problem

$$y = G(y_k) + \nabla G(y_k)'(y - y_k).$$

Thus,

$$y_{k+1} = G(y_k) + \nabla G(y_k)'(y_{k+1} - y_k),$$

or, by subtracting y_k from both sides and collecting terms,

$$\bigl(I - \nabla G(y_k)'\bigr)(y_{k+1} - y_k) = G(y_k) - y_k.$$

By using $H(y) = G(y) - y$, this equation can be written in the form of the Newton iteration (A.1).

Sec. A.1 Newton's Method for Differentiable Fixed Point Problems

By multiplying this relation with $\left(\nabla H(y_k)'\right)^{-1}$ we have

$$y_k - y^* - \left(\nabla H(y_k)'\right)^{-1} H(y_k) = o(\|y_k - y^*\|),$$

so for the Newton iteration (A.1), we obtain

$$y_{k+1} - y^* = o(\|y_k - y^*\|).$$

Thus, if $y_k \neq y^*$ for all k,

$$\lim_{k \to \infty} \frac{\|y_{k+1} - y^*\|}{\|y_k - y^*\|} = \lim_{k \to \infty} \frac{o(\|y_k - y^*\|)}{\|y_k - y^*\|} = 0,$$

implying superlinear convergence. This argument can also be used to show convergence to y^* if the initial vector y_0 is sufficiently close to y^*.

We will prove a more detailed version of this result, which includes a local convergence assertion ($\{y_k\}$ converges to y^* when started near y^*).

Proposition A.1: Consider a function $H : \Re^n \mapsto \Re^n$, and a vector y^* such that $H(y^*) = 0$. For any $\delta > 0$, we denote by S_δ the open sphere $\{x \mid \|y - y^*\| < \delta\}$, where $\|\cdot\|$ denotes the Euclidean norm. Assume that within some sphere $S_{\bar{\delta}}$, H is continuously differentiable, $\nabla H(y^*)$ is invertible, and $\left\|\left(\nabla H(y)'\right)^{-1}\right\|$ is bounded by some scalar $B > 0$:

$$\left\|\left(\nabla H(y)'\right)^{-1}\right\| \leq B, \qquad \text{for all } y \in S_{\bar{\delta}}.$$

Assume also that for some $L > 0$,

$$\|\nabla H(x) - \nabla H(y)\| \leq L \|x - y\|, \qquad \text{for all } x, y \in S_{\bar{\delta}},$$

Then there exists $\delta \in (0, \bar{\delta}]$ such that if $y_0 \in S_\delta$, the sequence $\{y_k\}$ generated by the iteration

$$y_{k+1} = y_k - \left(\nabla H(y_k)'\right)^{-1} H(y_k)$$

belongs to S_δ, and converges monotonically to y^*, i.e.

$$\|y_k - y^*\| \to 0, \qquad \|y_{k+1} - y^*\| \leq \|y_k - y^*\|, \qquad k = 0, 1, \ldots. \qquad (A.2)$$

Moreover, we have

$$\|y_{k+1} - y^*\| \leq \frac{LB}{2} \|y_k - y^*\|^2, \qquad k = 0, 1, \ldots. \qquad (A.3)$$

Proof: We first note that if $y_k \in S_{\bar{\delta}}$, by using the relation

$$H(y_k) = \int_0^1 \nabla H(y^* + t(y_k - y^*))' dt (y_k - y^*),$$

we have

$$\|y_{k+1} - y^*\| = \left\|y_k - y^* - \left(\nabla H(y_k)'\right)^{-1} H(y_k)\right\|$$

$$= \left\|\left(\nabla H(y_k)'\right)^{-1} \left(\nabla H(y_k)'(y_k - y^*) - H(y_k)\right)\right\|$$

$$= \left\|\left(\nabla H(y_k)'\right)^{-1} \left(\nabla H(y_k)' - \int_0^1 \nabla H(y^* + t(y_k - y^*))' dt\right) (y_k - y^*)\right\|$$

$$= \left\|\left(\nabla H(y_k)'\right)^{-1} \left(\int_0^1 \left[\nabla H(y_k)' - \nabla H(y^* + t(y_k - y^*))'\right] dt\right) (y_k - y^*)\right\|$$

$$\leq B \left(\int_0^1 \left\|\nabla H(y_k) - \nabla H(y^* + t(y_k - y^*))\right\| dt\right) \|y_k - y^*\|$$

$$\leq B \left(\int_0^1 Lt \|y_k - y^*\| dt\right) \|y_k - y^*\|$$

$$= \frac{LB}{2} \|y_k - y^*\|^2,$$

thus showing Eq. (A.3). Assume that $y_0 \in S_{\bar{\delta}}$. By continuity of ∇H, we can take $\delta \in (0, \bar{\delta}]$ such that $LB\delta < 1$, so if $y_0 \in S_\delta$, from the preceding relation we obtain

$$\|y_1 - y^*\| \leq \frac{1}{2} \|y_0 - y^*\| < \frac{\delta}{2}.$$

By repeating this argument with y_1 in place of y_0, we obtain $\|y_2 - y^*\| \leq \frac{1}{2} \|y_1 - y^*\| < \frac{\delta}{4}$, and similarly

$$\|y_{k+1} - y^*\| \leq \frac{1}{2} \|y_k - y^*\| < \frac{\delta}{2^{k+1}}, \qquad k = 0, 1, \dots.$$

The monotonic convergence property (A.2) follows. **Q.E.D.**

A.2 NEWTON'S METHOD WITHOUT DIFFERENTIABILITY OF THE BELLMAN OPERATOR

As we noted in Chapter 3, there is considerable literature on extensions of Newton's method for solution of the fixed point problem $y = G(y)$, which relax the differentiability requirement on G by using alternative notions from nonsmooth analysis. Relevant works include Josephy [Jos79], Robinson [Rob80], [Rob88], [Rob11], Kojima and Shindo [KoS86], Kummer [Kum88], [Kum00], Pang [Pan90], Qi and Sun [Qi93], [QiS93], Facchinei and Pang [FaP03], Ito and Kunisch [ItK03], Bolte, Daniilidis, and Lewis [BDL09], Dontchev and Rockafellar [DoR14], and additional references cited therein.

The Newton method extensions in these works have strong relevance to our context. In particular, the proof of Prop. A.1 can be simply extended to the case where G is nondifferentiable and has the minimization structure of the Bellman operator (assuming a finite control space). The idea is that when the kth iterate y_k is sufficiently close to the fixed point y^* of G, the kth Newton iteration can be viewed as a Newton iteration for *some* continuously differentiable mapping $\hat{G}_k$, which also has y^* as fixed point, and is obtained from G by a minimization operation. The preceding Prop. A.1, applied to $\hat{G}_k$, shows then that the distance $\|y_k - y^*\|$ decreases monotonically at a quadratic rate.

In the nondifferentiable case discussed in this appendix, we focus on solution of the equation $H(y) = 0$ where the mapping $H : \Re^n \mapsto \Re^n$ again has real-valued n components, denoted by $H_1(y), \ldots, H_n(y)$:

$$H(y) = \begin{pmatrix} H_1(y) \\ \vdots \\ H_n(y) \end{pmatrix}.$$

The component mappings $H_1, \ldots, H_n : \Re^n \to \Re$, involve minimization over a parameter u (as the notation suggests, the parameter corresponds to control in the DP context), and have the form

$$H_i(y) = \min_{u=1,\ldots,m} H_{i,u}(y), \qquad i = 1, \ldots, n. \tag{A.4}$$

The mappings $H_{i,u} : \Re^n \to \Re$ are given by

$$H_{i,u}(y) = G_{i,u}(y) - y_i, \qquad i = 1, \ldots, n, \ u = 1, \ldots, m,$$

where $G_{i,u} : \Re^n \mapsto \Re$ is a given real-valued function for each i and u, and y_i is the ith component of y. Given a vector y^* such that $H(y^*) = 0$, we denote by $U^*(i) \subset \{1, \ldots, m\}$ the set of indexes that attain the minimum in Eq. (A.4) when $y = y^*$:

$$U^*(i) = \arg\min_{u=1,\ldots,m} H_{i,u}(y^*), \qquad i = 1, \ldots, n.$$

We assume that within some sphere $S_{\bar{\delta}}$ centered at y^* with radius $\bar{\delta}$, the mappings $H_{i,u}(\cdot)$ are continuously differentiable for all i and $u \in U^*(i)$, while all the $n \times n$ matrices with columns

$$\nabla H_{1,u_1}(y), \ldots, \nabla H_{n,u_n}(y),$$

are invertible, where for every i, u_i can take any value from the set $U^*(i)$. Thus all the Jacobian matrices of the mappings, which correspond to $(u_1, \ldots, u_n)$ that are "active" at y^* [i.e., $u_i \in U^*(i)$ for all i], are assumed invertible.

Given the iterate y_k, Newton's method operates as follows: It finds for each $i = 1, \ldots, n$, the set of indexes $U(i,k) \subset \{1, \ldots, m\}$ that attain the minimum in Eq. (A.4) when $y = y_k$:

$$U(i,k) = \arg \min_{u=1,\ldots,m} H_{i,u}(y_k).$$

Then it generates the next iterate y_{k+1} with the following three steps:

(a) It selects arbitrarily an index $u(i,k)$ from within $U(i,k)$, for each $i = 1, \ldots, n$.

(b) It forms the $n \times n$ matrix

$$M_k = \big(\nabla H_{1,u(1,k)}(y_k) \;\cdots\; \nabla H_{n,u(n,k)}(y_k)\big),$$

that has columns $\nabla H_{1,u(1,k)}(y_k), \ldots, \nabla H_{n,u(n,k)}(y_k)$, and the column vector

$$G_k = \begin{pmatrix} H_{1,u(1,k)}(y_k) \\ \vdots \\ H_{n,u(n,k)}(y_k) \end{pmatrix}$$

that has components $H_{1,u(1,k)}(y_k), \ldots, H_{n,u(n,k)}(y_k)$:

(c) It sets

$$y_{k+1} = y_k - (M'_k)^{-1} G_k. \tag{A.5}$$

For our convergence proof, we argue as follows: When the iterate y_k is sufficiently near to y^*, the index set $U(i,k)$ is a subset of $U^*(i)$ for each $i = 1, \ldots, n$; see Fig. A.2 for a case where $n = 1$ and $m = 3$. The reason is that there exists an $\epsilon > 0$ such that for all $i = 1, \ldots, n$ and $u = 1, \ldots, m$,

$$u \notin U^*(i) \quad \Rightarrow \quad H_{i,u}(y^*) \geq \epsilon.$$

Therefore, there is a sphere centered at y^* such that for all y_k within that sphere and all i, we have

$$u \notin U^*(i) \quad \Rightarrow \quad H_{i,u}(y_k) \geq \epsilon/2,$$

Sec. A.2 Newton's Method Without Differentiability

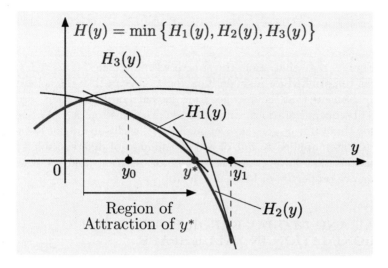

Figure A.2 Illustration of a one-dimensional nondifferentiable equation

$$H(y) = 0,$$

where H is obtained by minimization of three differentiable mappings H_1, H_2, H_3:

$$H(y) = \min\{H_1(y), H_2(y), H_3(y)\}, \qquad y \in \Re.$$

Here $m = 3$ and $n = 1$ [so the index i is omitted from $U^*(i)$ and $U(i,k)$]. At the solution y^*, the index set U^* consists of $u = 1$ and $u = 2$, and for $y_k \neq y^*$, we have $U(k) = \{1\}$ or $U(k) = \{2\}$. Newton's method applied to H consists of a Newton iteration applied to H_1 when $y < y^*$ and $|y - y^*|$ is small enough, and consists of a Newton iteration applied to H_2 when $y > y^*$. Since at y^* we have

$$H_1(y^*) = H_2(y^*) = 0,$$

both iterations, at $y < y^*$ and at $y > y^*$, approach y^* superlinearly.

and
$$u \in U^*(i) \quad \Rightarrow \quad H_{i,u}(y_k) < \epsilon/2,$$

which implies that if $u \notin U^*(i)$ then $u \notin U(i,k)$, or equivalently $U(i,k) \subset U^*(i)$. It follows that *the iteration (A.5) can be viewed as a Newton iteration applied to a system of differentiable equations that has y^* as its solution.* This system is

$$H_{i,u(i,k)}(y) = 0, \qquad i = 1, \ldots, n,$$

and corresponds to the set of indexes

$$u(1,k), \ldots, u(n,k).$$

Thus, near y^*, by Prop. A.1, the iteration (A.5) is attracted at a quadratic rate to y^* regardless of which indexes $u(i,k) \in U(i,k)$ are selected for iteration k.

Finally, note that while the sphere within which $U(i,k) \subset U^*(i)$ for all i was constructed for a single iteration k, we can take the sphere small enough to ensure that the distance from the current iterate to y^* is reduced for all subsequent iterations, similar to the proof of Prop. A.1. We can thus conclude that after y_k gets close enough to y^*, each subsequent iteration is a Newton step applied to *one* of a finite number of differentiable systems of equations that have y^* as their common solution, and for which the convergence properties of Prop. A.1 hold.

A.3 LOCAL AND GLOBAL ERROR BOUNDS FOR APPROXIMATION IN VALUE SPACE

In approximation in value space, an important analytical issue is to quantify the level of suboptimality of the one-step or multistep lookahead policy obtained. In this section, we focus on an ℓ-step lookahead scheme that produces a policy $\tilde{\mu}$ according to

$$T_{\tilde{\mu}} T^{\ell-1} \tilde{J} = T^\ell \tilde{J},$$

where $\tilde{J}$ is the terminal cost function approximation. We will try to estimate the difference $J_{\tilde{\mu}} - J^*$, where $J_{\tilde{\mu}}$ is the cost function of $\tilde{\mu}$ and J^* is the optimal cost function, assuming that $T^{\ell-1}\tilde{J}$ lies within the region of stability, so that $\tilde{\mu}$ is well-defined as a stable policy and $J_{\tilde{\mu}}$ is finite-valued.

There is a classical error bound for the case where the Bellman operator T is a contraction mapping with respect to the sup-norm ($\|J\| = \sup_{x \in X} |J(x)|$) with modulus $\alpha \in (0,1)$. It is given by

$$\|J_{\tilde{\mu}} - J^*\| \le \frac{2\alpha}{1-\alpha} \|T^{\ell-1}\tilde{J} - J^*\|; \tag{A.6}$$

see [Ber19a], Prop. 5.1.1, for the finite-state discounted case, and [Ber22a], Section 2.2, for more general abstract DP cases. This bound also applies to the case where T is a contraction over a subset $\mathcal{J}$ of functions, as long as T maps $\mathcal{J}$ into itself.

Unfortunately, however, this error bound is very conservative, and does not reflect practical reality. The reason is that this is a *global* error bound, i.e., it holds for all $\tilde{J}$, even the worst possible. In practice, $\tilde{J}$ is often chosen sufficiently close to J^*, so that the error $J_{\tilde{\mu}} - J^*$ behaves consistently with the convergence rate of the Newton step that starts at $T^{\ell-1}\tilde{J}$, which is superlinear. In other words, for $\tilde{J}$ relatively close to J^*, we have the *local* estimate

$$\|J_{\tilde{\mu}} - J^*\| = o\bigl(\|T^{\ell-1}\tilde{J} - J^*\|\bigr). \tag{A.7}$$

Sec. A.3   Error Bounds for Approximation in Value Space

In practical terms, there is often a huge difference, both quantitative and qualitative, between the error bounds (A.6) and (A.7), as illustrated by the following example.

Example A.3.1 (One-Dimensional Linear Quadratic Problem)

Consider an undiscounted one-dimensional linear quadratic problem such as the one considered in Chapter 4. The system is

$$x_{k+1} = ax_k + bu_k,$$

and the cost per stage is

$$qx_k^2 + ru_k^2.$$

We will consider one-step lookahead ($\ell = 1$), and a quadratic cost function approximation

$$\tilde{J}(x) = \tilde{K}x^2,$$

with $\tilde{K}$ within the region of stability, which is some open interval of the form (S, ∞). As in Chapter 4, the Riccati operator is

$$F(K) = \frac{a^2 rK}{r + b^2 K} + q,$$

and the one-step lookahead policy $\tilde{\mu}$ has cost function

$$J_{\tilde{\mu}}(x) = K_{\tilde{\mu}} x^2,$$

where $K_{\tilde{\mu}}$ is obtained by applying one step of Newton's method for solving the Riccati equation $K = F(K)$, starting at $K = \tilde{K}$.

Let S be the boundary of the region of stability, i.e., the value of K at which the derivative of F with respect to K is equal to 1:

$$\left.\frac{\partial F(K)}{\partial K}\right|_{K=S} = 1.$$

Then the Riccati operator F is a contraction within any interval $[\overline{S}, \infty)$ with $\overline{S} > S$, with a contraction modulus α that depends on $\overline{S}$. In particular, α is given by

$$\alpha = \left.\frac{\partial F(K)}{\partial K}\right|_{K=\overline{S}}$$

and satisfies $0 < \alpha < 1$ because $\overline{S} > S$, and the derivative of F is positive and monotonically decreasing to 0 as K increases to ∞.

The error bound (A.6) can be rederived for the case of quadratic functions and can be rewritten in terms of quadratic cost coefficients as

$$K_{\tilde{\mu}} - K^* \leq \frac{2\alpha}{1-\alpha}|\tilde{K} - K^*|, \tag{A.8}$$

where $K_{\tilde{\mu}}$ is the quadratic cost coefficient of the lookahead policy $\tilde{\mu}$ [and also the result of a Newton step for solving the fixed point Riccati equation

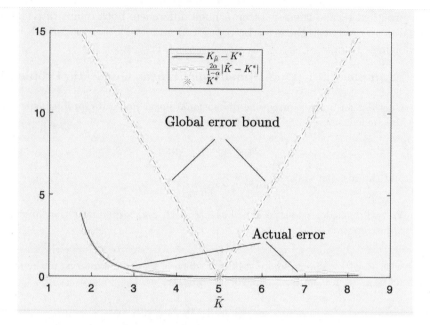

Figure A.3 Illustration of the global error bound (A.8) for the one step lookahead error $K_{\tilde\mu} - K^*$ as a function of $\tilde K$, compared with the true error obtained by one step of Newton's method starting from $\tilde K$.

The problem data are $a = 2$, $b = 2$, $q = 1$, and $r = 5$. With these numerical values, we have $K^* = 5$ and the region of stability is (S, ∞) with $S = 1.25$. The modulus of contraction α used in the figure is computed at $\overline S = S + 0.5$. Depending on the chosen value of $\overline S$, α can be arbitrarily close to 1, but decreases as $\overline S$ increases. Note that the error $K_{\tilde\mu} - K^*$ is much smaller when $\tilde K$ is larger than K^* than when it is lower, because the slope of F diminishes as K increases. This is not reflected by the global error bound (A.8).

$F = F(K)$ starting from $\tilde K$]. A plot of $(K_{\tilde\mu} - K^*)$ as a function of $\tilde K$, compared with the bound on the right side of this equation is shown in Fig. A.3. It can be seen that $(K_{\tilde\mu} - K^*)$ exhibits the qualitative behavior of Newton's method, which is very different than the bound (A.8). An interesting fact is that the bound (A.8) depends on α, which in turn depends on how close $\tilde K$ is to the boundary S of the region of stability, while the local behavior of Newton's method is independent of S.

A.4 LOCAL AND GLOBAL ERROR BOUNDS FOR APPROXIMATE POLICY ITERATION

In an approximate PI method that generates a sequence of policies $\{\mu^k\}$, it is important to estimate the asymptotic error

$$\limsup_{k\to\infty} \|J_{\mu^k} - J^*\|.$$

Sec. A.4   Error Bounds for Approximate Policy Iteration

In this section we will consider the case where the policy evaluation step is approximate, while the policy improvement step is exact (so that it is equivalent to a Newton step for solving the Bellman equation $J = TJ$). In particular, we focus on a method that generates a sequence of policies $\{\mu^k\}$ and a corresponding sequence of approximate cost functions $\{J_k\}$ satisfying

$$T_{\mu^{k+1}} J_k = T J_k, \qquad \|J_{k+1} - J_{\mu^{k+1}}\| \leq \delta, \qquad k = 0, 1, \ldots,$$

for some $\delta > 0$. Here J_0 is some initial cost function approximation, μ^1 is the first policy, obtained from J_0 by a one-step lookahead/Newton step, and δ is some scalar quantifying the level of approximation in the policy evaluation step (i.e., replacing J_{μ^1} with J_1, and more generally replacing $J_{\mu^{k+1}}$ with J_{k+1}).

A prerequisite for this method to be well defined is that the generated sequence $\{J_k\}$ stays within the region of stability of the problem, since otherwise the policy improvement step, $T_{\mu^{k+1}} J_k = T J_k$, is not well-defined. This difficulty does not arise in problems where T is a contraction mapping, such as in discounted problems with bounded cost per stage, but suggests that in general the size of δ should be small enough to bring J_k relatively close to J^*, and keep it close.

For the remainder of this section, we assume that T is a contraction mapping with modulus $\alpha \in (0,1)$. The main global error bound for approximate PI under this assumption is

$$\limsup_{k \to \infty} \|J_{\mu^k} - J^*\| \leq \frac{2\alpha\delta}{(1-\alpha)^2}. \tag{A.9}$$

It was first given in the book [BeT96] for finite-state discounted problems, and a proof that applies to general abstract DP mappings was given in the book [Ber22a], Section 2.4.1. The essence of the proof is the following inequality, which quantifies the amount of approximate policy improvement at each iteration:

$$\|J_{\mu^{k+1}} - J^*\| \leq \alpha \|J_{\mu^k} - J^*\| + \frac{2\alpha\delta}{1-\alpha}.$$

It states that the error is reduced geometrically by a factor α, plus an additive constant term $\frac{2\alpha\delta}{1-\alpha}$, which over time accumulates to $\frac{2\alpha\delta}{(1-\alpha)^2}$.

Unfortunately, the error bound (A.9) is very conservative, and does not reflect practical reality. In applications of approximate PI, the iterates J_k often get sufficiently close to J^*, for the error $J_{\mu^{k+1}} - J^*$ to have the superlinear convergence rate of Newton's method:

$$\|J_{\mu^{k+1}} - J^*\| = o(\|J_k - J^*\|).$$

This suggests that $J_{\mu^{k+1}} - J^*$ converges to a neighborhood of J^* of size $O(\delta)$ once $\|J_k - J^*\|$ becomes small. An extreme manifestation of this arises when the number of policies is finite, in which case J^* is piecewise linear, as in the case of a finite-spaces α-discounted problem. Then it can be shown that once $\|J_k - J^*\|$ becomes sufficiently small, approximate policy iteration produces an optimal policy at the next iteration, a property not captured by the global error bound (A.9). A related property is that when the number of policies is finite and δ is sufficiently small, then approximate policy iteration produces an optimal policy in a finite number of iterations, something that is also not captured by Eq. (A.9).

For another view of this phenomenon, let us use the triangle inequality and the definition of δ as a bound of the policy evaluation error to write

$$\|J_{k+1} - J^*\| \leq \|J_{\mu^{k+1}} - J^*\| + \|J_{k+1} - J_{\mu^{k+1}}\| \leq \|J_{\mu^{k+1}} - J^*\| + \delta,$$

which in view of the superlinear convergence rate of Newton's method, $\|J_{\mu^{k+1}} - J^*\| = o(\|J_k - J^*\|)$, yields

$$\|J_{k+1} - J^*\| \leq o(\|J_k - J^*\|) + \delta. \tag{A.10}$$

This relation suggests again that once J_k gets within a neighborhood of size that is comparable to δ, it tends to stay within that neighborhood. As an indication of this, note that when T is linear, then the $o(\|J_k - J^*\|)$ term in Eq. (A.10) is equal to 0, so the error bound becomes

$$\|J_{k+1} - J^*\| \leq \delta,$$

regardless of the modulus of contraction α. Moreover, the same is true when T is piecewise linear and δ is sufficiently small. We can also argue that a similar behavior occurs within a small neighborhood of J^* when δ is small, but a detailed analysis will not be presented here. Instead, we will illustrate the behavior of the error $J_{\mu^{k+1}} - J^*$ for linear quadratic problems.

Example A.4.1 (Approximate PI for a One-Dimensional Linear Quadratic Problem)

We consider the linear quadratic problem of Example A.3.1, and the approximate PI algorithm with approximate policy evaluation within δ, in the sense that

$$|K_k - K_{\mu^k}| \leq \delta.$$

We will compute the asymptotic error

$$\limsup_{k \to \infty} (K_{\mu^k} - K^*), \tag{A.11}$$

as a function of δ, and we will then compare it to the version of the global error bound (A.9) for linear quadratic problems, which takes the form

$$\limsup_{k \to \infty} (K_{\mu^k} - K^*) \leq \frac{2\alpha\delta}{(1-\alpha)^2}, \tag{A.12}$$

with α being a suitable contraction modulus.

Figure A.4 provides the comparison for the same problem data, $a = 2$, $b = 2$, $q = 1$, and $r = 5$, as in Fig. A.3. Note that the true error is roughly equal to δ as suggested by the discussion that precedes the present example.

Sec. A.4 Error Bounds for Approximate Policy Iteration

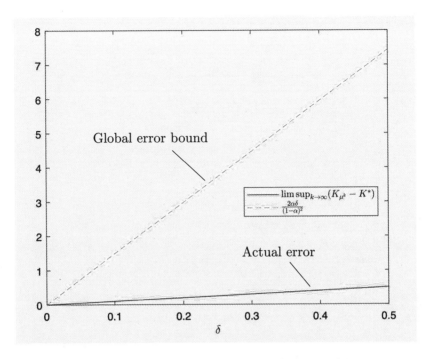

Figure A.4 Illustration of the asymptotic global error bound (A.9) for PI with approximate policy evaluation as a function of δ, compared with the true asymptotic error.

References

[ADB17] Arulkumaran, K., Deisenroth, M. P., Brundage, M., and Bharath, A. A., 2017. "A Brief Survey of Deep Reinforcement Learning," arXiv preprint arXiv:1708.05866.

[Arg08] Argyros, I. K., 2008. Convergence and Applications of Newton-Type Iterations, Springer, N. Y.

[AsH95] Aström, K. J., and Hagglund, T., 1995. PID Controllers: Theory, Design, and Tuning, Instrument Society of America, Research Triangle Park, NC.

[AsH06] Aström, K. J., and Hagglund, T., 2006. Advanced PID Control, Instrument Society of America, Research Triangle Park, N. C.

[AsW08] Aström, K. J., and Wittenmark, B., 2008. Adaptive Control, Dover Books; also Prentice-Hall, Englewood Cliffs, N. J, 1994.

[BBB22] Bhambri, S., Bhattacharjee, A., and Bertsekas, D. P., 2022. "Reinforcement Learning Methods for Wordle: A POMDP/Adaptive Control Approach," arXiv preprint arXiv:2211.10298.

[BBD10] Busoniu, L., Babuska, R., De Schutter, B., and Ernst, D., 2010. Reinforcement Learning and Dynamic Programming Using Function Approximators, CRC Press, N. Y.

[BBM17] Borrelli, F., Bemporad, A., and Morari, M., 2017. Predictive Control for Linear and Hybrid Systems, Cambridge Univ. Press, Cambridge, UK.

[BBS95] Barto, A. G., Bradtke, S. J., and Singh, S. P., 1995. "Real-Time Learning and Control Using Asynchronous Dynamic Programming," Artificial Intelligence, Vol. 72, pp. 81-138.

[BDL09] Bolte, J., Daniilidis, A., and Lewis, A., 2009. "Tame Functions are Semismooth," Math. Programming, Vol. 117, pp. 5-19.

[BDT18] Busoniu, L., de Bruin, T., Tolic, D., Kober, J., and Palunko, I., 2018. "Reinforcement Learning for Control: Performance, Stability, and Deep Approximators," Annual Reviews in Control, Vol. 46, pp. 8-28.

[BKB20] Bhattacharya, S., Kailas, S., Badyal, S., Gil, S., and Bertsekas, D. P., 2020. "Multiagent Rollout and Policy Iteration for POMDP with Application to Multi-Robot Repair Problems," in Proc. of Conference on Robot Learning (CoRL); also arXiv preprint, arXiv:2011.04222.

[BLW91] Bittanti, S., Laub, A. J., and Willems, J. C., eds., 2012. The Riccati Equation, Springer.

[BMZ09] Bokanowski, O., Maroso, S., and Zidani, H., 2009. "Some Convergence Results for Howard's Algorithm," SIAM J. on Numerical Analysis, Vol. 47, pp. 3001-3026.

[BPW12] Browne, C., Powley, E., Whitehouse, D., Lucas, L., Cowling, P. I., Rohlfshagen, P., Tavener, S., Perez, D., Samothrakis, S., and Colton, S., 2012. "A Survey of Monte Carlo Tree Search Methods," IEEE Trans. on Computational Intelligence and AI in Games, Vol. 4, pp. 1-43.

[BSW99] Beard, R. W., Saridis, G. N., and Wen, J. T., 1998. "Approximate Solutions to the Time-Invariant Hamilton-Jacobi-Bellman Equation," J. of Optimization Theory and Applications, Vol. 96, pp. 589-626.

[BTW97] Bertsekas, D. P., Tsitsiklis, J. N., and Wu, C., 1997. "Rollout Algorithms for Combinatorial Optimization," Heuristics, Vol. 3, pp. 245-262.

[BeC99] Bertsekas, D. P., and Castañon, D. A., 1999. "Rollout Algorithms for Stochastic Scheduling Problems," Heuristics, Vol. 5, pp. 89-108.

[BeI96] Bertsekas, D. P., and Ioffe, S., 1996. "Temporal Differences-Based Policy Iteration and Applications in Neuro-Dynamic Programming," Lab. for Info. and Decision Systems Report LIDS-P-2349, MIT, Cambridge, MA.

[BeK65] Bellman, R., and Kalaba, R. E., 1965. Quasilinearization and Nonlinear Boundary-Value Problems, Elsevier, N.Y.

[BeR71] Bertsekas, D. P., and Rhodes, I. B., 1971. "On the Minimax Reachability of Target Sets and Target Tubes," Automatica, Vol. 7, pp. 233-247.

[BeS78] Bertsekas, D. P., and Shreve, S. E., 1978. Stochastic Optimal Control: The Discrete Time Case, Academic Press, N. Y.; republished by Athena Scientific, Belmont, MA, 1996 (can be downloaded from the author's website).

[BeT89] Bertsekas, D. P., and Tsitsiklis, J. N., 1989. Parallel and Distributed Computation: Numerical Methods, Prentice-Hall, Engl. Cliffs, N. J. (can be downloaded from the author's website).

[BeT96] Bertsekas, D. P., and Tsitsiklis, J. N., 1996. Neuro-Dynamic Programming, Athena Scientific, Belmont, MA.

[BeT08] Bertsekas, D. P., and Tsitsiklis, J. N., 2008. Introduction to Probability, 2nd Edition, Athena Scientific, Belmont, MA.

[BeY10] Bertsekas, D. P., and Yu, H., 2010. "Distributed Asynchronous Policy Iteration in Dynamic Programming," Proc. of Allerton Conf. on Communication, Control and Computing, Allerton Park, Ill, pp. 1368-1374.

[BeY12] Bertsekas, D. P., and Yu, H., 2012. "Q-Learning and Enhanced Policy Iteration in Discounted Dynamic Programming," Math. of Operations Research, Vol. 37, pp. 66-94.

[BeY16] Bertsekas, D. P., and Yu, H., 2016. "Stochastic Shortest Path Problems Under Weak Conditions," Lab. for Information and Decision Systems Report LIDS-2909, Massachusetts Institute of Technology.

[Bea95] Beard, R. W., 1995. Improving the Closed-Loop Performance of Nonlinear Systems, Ph.D. Thesis, Rensselaer Polytechnic Institute.

[Bel57] Bellman, R., 1957. Dynamic Programming, Princeton University Press, Princeton, N. J.

[Ber71] Bertsekas, D. P., 1971. "Control of Uncertain Systems With a Set-Membership Description of the Uncertainty," Ph.D. Thesis, Massachusetts Institute of

References

Technology, Cambridge, MA (can be downloaded from the author's website).

[Ber72] Bertsekas, D. P., 1972. "Infinite Time Reachability of State Space Regions by Using Feedback Control," IEEE Trans. Automatic Control, Vol. AC-17, pp. 604-613.

[Ber82] Bertsekas, D. P., 1982. "Distributed Dynamic Programming," IEEE Trans. Aut. Control, Vol. AC-27, pp. 610-616.

[Ber83] Bertsekas, D. P., 1983. "Asynchronous Distributed Computation of Fixed Points," Math. Programming, Vol. 27, pp. 107-120.

[Ber05] Bertsekas, D. P., 2005. "Dynamic Programming and Suboptimal Control: A Survey from ADP to MPC," European J. of Control, Vol. 11, pp. 310-334.

[Ber11] Bertsekas, D. P., 2011. "Approximate Policy Iteration: A Survey and Some New Methods," J. of Control Theory and Applications, Vol. 9, pp. 310-335.

[Ber12] Bertsekas, D. P., 2012. Dynamic Programming and Optimal Control, Vol. II, 4th Ed., Athena Scientific, Belmont, MA.

[Ber15] Bertsekas, D. P., 2015. "Lambda-Policy Iteration: A Review and a New Implementation," arXiv preprint arXiv:1507.01029.

[Ber16] Bertsekas, D. P., 2016. Nonlinear Programming, Athena Scientific, Belmont, MA.

[Ber17a] Bertsekas, D. P., 2017. Dynamic Programming and Optimal Control, Vol. I, 4th Ed., Athena Scientific, Belmont, MA.

[Ber17b] Bertsekas, D. P., 2017. "Value and Policy Iteration in Deterministic Optimal Control and Adaptive Dynamic Programming," IEEE Trans. on Neural Networks and Learning Systems, Vol. 28, pp. 500-509.

[Ber17c] Bertsekas, D. P., 2017. "Regular Policies in Abstract Dynamic Programming," SIAM J. on Optimization, Vol. 27, pp. 1694-1727.

[Ber18a] Bertsekas, D. P., 2018. Abstract Dynamic Programming, 2nd Ed., Athena Scientific, Belmont, MA (can be downloaded from the author's website).

[Ber18b] Bertsekas, D. P., 2018. "Feature-Based Aggregation and Deep Reinforcement Learning: A Survey and Some New Implementations," Lab. for Information and Decision Systems Report, MIT; arXiv preprint arXiv:1804.04577; IEEE/CAA Journal of Automatica Sinica, Vol. 6, 2019, pp. 1-31.

[Ber18c] Bertsekas, D. P., 2018. "Biased Aggregation, Rollout, and Enhanced Policy Improvement for Reinforcement Learning," Lab. for Information and Decision Systems Report, MIT; arXiv preprint arXiv:1910.02426.

[Ber18d] Bertsekas, D. P., 2018. "Proximal Algorithms and Temporal Difference Methods for Solving Fixed Point Problems," Computational Optimization and Applications, Vol. 70, pp. 709-736.

[Ber19a] Bertsekas, D. P., 2019. Reinforcement Learning and Optimal Control, Athena Scientific, Belmont, MA.

[Ber19b] Bertsekas, D. P., 2019. "Multiagent Rollout Algorithms and Reinforcement Learning," arXiv preprint arXiv:1910.00120.

[Ber19c] Bertsekas, D. P., 2019. "Constrained Multiagent Rollout and Multidimensional Assignment with the Auction Algorithm," arXiv preprint, arxiv.org/abs/2002.07407.

[Ber20a] Bertsekas, D. P., 2020. Rollout, Policy Iteration, and Distributed Reinforcement Learning, Athena Scientific, Belmont, MA.

[Ber20b] Bertsekas, D. P., 2020. "Multiagent Value Iteration Algorithms in Dynamic Programming and Reinforcement Learning," Results in Control and Optimization J., Vol. 1, 2020.

[Ber21a] Bertsekas, D. P., 2021. "On-Line Policy Iteration for Infinite Horizon Dynamic Programming," arXiv preprint arXiv:2106.00746.

[Ber21b] Bertsekas, D. P., 2021. "Multiagent Reinforcement Learning: Rollout and Policy Iteration," IEEE/ CAA J. of Automatica Sinica, Vol. 8, pp. 249-271.

[Ber21c] Bertsekas, D. P., 2021. "Distributed Asynchronous Policy Iteration for Sequential Zero-Sum Games and Minimax Control," arXiv preprint arXiv:2107.-10406.

[Ber22a] Bertsekas, D. P., 2022. Abstract Dynamic Programming, 3rd Edition, Athena Scientific, Belmont, MA (can be downloaded from the author's website).

[Ber22b] Bertsekas, D. P., 2022. "Newton's Method for Reinforcement Learning and Model Predictive Control," Results in Control and Optimization J., Vol. 7, 2022, pp. 100-121.

[Ber22c] Bertsekas, D. P., 2022. "Rollout Algorithms and Approximate Dynamic Programming for Bayesian Optimization and Sequential Estimation," arXiv preprint arXiv:2212.07998v3.

[Bit91] Bittanti, S., 1991. "Count Riccati and the Early Days of the Riccati Equation," in The Riccati Equation (pp. 1-10), Springer.

[Bla65] Blackwell, D., 1965. "Positive Dynamic Programming," Proc. Fifth Berkeley Symposium Math. Statistics and Probability, pp. 415-418.

[BoV79] Borkar, V., and Varaiya, P. P., 1979. "Adaptive Control of Markov Chains, I: Finite Parameter Set," IEEE Trans. Automatic Control, Vol. AC-24, pp. 953-958.

[Bod20] Bodson, M., 2020. Adaptive Estimation and Control, Independently Published.

[Bor08] Borkar, V. S., 2008. Stochastic Approximation: A Dynamical Systems Viewpoint, Cambridge Univ. Press.

[Bor09] Borkar, V. S., 2009. "Reinforcement Learning: A Bridge Between Numerical Methods and Monte Carlo," in World Scientific Review, Vol. 9, Ch. 4.

[Bra21] Brandimarte, P., 2021. From Shortest Paths to Reinforcement Learning: A MATLAB-Based Tutorial on Dynamic Programming, Springer.

[CFH13] Chang, H. S., Hu, J., Fu, M. C., and Marcus, S. I., 2013. Simulation-Based Algorithms for Markov Decision Processes, 2nd Edition, Springer, N. Y.

[CaB07] Camacho, E. F., and Bordons, C., 2007. Model Predictive Control, 2nd Ed., Springer, New York, N. Y.

[Cao07] Cao, X. R., 2007. Stochastic Learning and Optimization: A Sensitivity-Based Approach, Springer, N. Y.

[DNP11] Deisenroth, M. P., Neumann, G., and Peters, J., 2011. "A Survey on Policy Search for Robotics," Foundations and Trends in Robotics, Vol. 2, pp. 1-142.

[DeF04] De Farias, D. P., 2004. "The Linear Programming Approach to Approxi-

References

mate Dynamic Programming," in Learning and Approximate Dynamic Programming, by J. Si, A. Barto, W. Powell, and D. Wunsch, (Eds.), IEEE Press, N. Y.

[DoS80] Doshi, B., and Shreve, S., 1980. "Strong Consistency of a Modified Maximum Likelihood Estimator for Controlled Markov Chains," J. of Applied Probability, Vol. 17, pp. 726-734.

[DoR14] Dontchev, A. L., and Rockafellar, R. T., 2014. Implicit Functions and Solution Mappings, 2nd Edition, Springer, N. Y.

[FaP03] Facchinei, F., and Pang, J.-S., 2003. Finite-Dimensional Variational Inequalities and Complementarity Problems, Vols I and II, Springer, N. Y.

[FeS04] Ferrari, S., and Stengel, R. F., 2004. "Model-Based Adaptive Critic Designs," in Learning and Approximate Dynamic Programming, by J. Si, A. Barto, W. Powell, and D. Wunsch, (Eds.), IEEE Press, N. Y.

[GBL12] Grondman, I., Busoniu, L., Lopes, G. A. D., and Babuska, R., 2012. "A Survey of Actor-Critic Reinforcement Learning: Standard and Natural Policy Gradients," IEEE Trans. on Systems, Man, and Cybernetics, Part C, Vol. 42, pp. 1291-1307.

[GSD06] Goodwin, G., Seron, M. M., and De Dona, J. A., 2006. Constrained Control and Estimation: An Optimisation Approach, Springer, N. Y.

[GoS84] Goodwin, G. C., and Sin, K. S. S., 1984. Adaptive Filtering, Prediction, and Control, Prentice-Hall, Englewood Cliffs, N. J.

[Gos15] Gosavi, A., 2015. Simulation-Based Optimization: Parametric Optimization Techniques and Reinforcement Learning, 2nd Edition, Springer, N. Y.

[HWL21] Ha, M., Wang, D., and Liu, D., 2021. "Offline and Online Adaptive Critic Control Designs With Stability Guarantee Through Value Iteration," IEEE Transactions on Cybernetics.

[HaR21] Hardt, M.. and Recht, B., 2021. Patterns, Predictions, and Actions: A Story About Machine Learning, arXiv preprint arXiv:2102.05242.

[Hay08] Haykin, S., 2008. Neural Networks and Learning Machines, 3rd Ed., Prentice-Hall, Englewood-Cliffs, N. J.

[Hew71] Hewer, G., 1971. "An Iterative Technique for the Computation of the Steady State Gains for the Discrete Optimal Regulator," IEEE Trans. on Automatic Control, Vol. 16, pp. 382-384.

[Hey17] Heydari, A., 2017. "Stability Analysis of Optimal Adaptive Control Under Value Iteration Using a Stabilizing Initial Policy," IEEE Trans. on Neural Networks and Learning Systems, Vol. 29, pp. 4522-4527.

[Hey18] Heydari, A., 2018. "Stability Analysis of Optimal Adaptive Control Using Value Iteration with Approximation Errors," IEEE Transactions on Automatic Control, Vol. 63, pp. 3119-3126.

[Hyl11] Hylla, T., 2011. Extension of Inexact Kleinman-Newton Methods to a General Monotonicity Preserving Convergence Theory, Ph.D. Thesis, Univ. of Trier.

[IoS96] Ioannou, P. A., and Sun, J., 1996. Robust Adaptive Control, Prentice-Hall, Englewood Cliffs, N. J.

[ItK03] Ito, K., and Kunisch, K., 2003. "Semi-Smooth Newton Methods for Variational Inequalities of the First Kind," Mathematical Modelling and Numerical Analysis, Vol. 37, pp. 41-62.

[JiJ17] Jiang, Y., and Jiang, Z. P., 2017. Robust Adaptive Dynamic Programming, J. Wiley, N. Y.

[Jos79] Josephy, N. H., 1979. "Newton's Method for Generalized Equations," Wisconsin Univ-Madison, Mathematics Research Center Report No. 1965.

[KAC15] Kochenderfer, M. J., with Amato, C., Chowdhary, G., How, J. P., Davison Reynolds, H. J., Thornton, J. R., Torres-Carrasquillo, P. A., Ore, N. K., Vian, J., 2015. Decision Making under Uncertainty: Theory and Application, MIT Press, Cambridge, MA.

[KKK95] Krstic, M., Kanellakopoulos, I., Kokotovic, P., 1995. Nonlinear and Adaptive Control Design, J. Wiley, N. Y.

[KKK20] Kalise, D., Kundu, S., and Kunisch, K., 2020. "Robust Feedback Control of Nonlinear PDEs by Numerical Approximation of High-Dimensional Hamilton-Jacobi-Isaacs Equations." SIAM J. on Applied Dynamical Systems, Vol. 19, pp. 1496-1524.

[KLM96] Kaelbling, L. P., Littman, M. L., and Moore, A. W., 1996. "Reinforcement Learning: A Survey," J. of Artificial Intelligence Res., Vol. 4, pp. 237-285.

[KeG88] Keerthi, S. S., and Gilbert, E. G., 1988. "Optimal Infinite-Horizon Feedback Laws for a General Class of Constrained Discrete-Time Systems: Stability and Moving-Horizon Approximations," J. Optimization Theory Appl., Vo. 57, pp. 265-293.

[Kle67] Kleinman, D. L., 1967. Suboptimal Design of Linear Regulator Systems Subject to Computer Storage Limitations, Doctoral dissertation, M.I.T., Electronic Systems Lab., Rept. 297.

[Kle68] Kleinman, D. L., 1968. "On an Iterative Technique for Riccati Equation Computations," IEEE Trans. Automatic Control, Vol. AC-13, pp. 114-115.

[KoC16] Kouvaritakis, B., and Cannon, M., 2016. Model Predictive Control: Classical, Robust and Stochastic, Springer, N. Y.

[KoS86] Kojima, M., and Shindo, S., 1986. "Extension of Newton and Quasi-Newton Methods to Systems of PC^1 Equations," J. of the Operations Research Society of Japan, Vol. 29, pp. 352-375.

[Kok91] Kokotovic, P. V., ed., 1991. Foundations of Adaptive Control, Springer.

[Kor90] Korf, R. E., 1990. "Real-Time Heuristic Search," Artificial Intelligence, Vol. 42, pp. 189-211.

[Kri16] Krishnamurthy, V., 2016. Partially Observed Markov Decision Processes, Cambridge Univ. Press.

[Kum88] Kummer, B., 1988. "Newton's Method for Non-Differentiable Functions," Mathematical Research, Vol. 45, pp. 114-125.

[Kum00] Kummer, B., 2000. "Generalized Newton and NCP-methods: Convergence, Regularity, Actions," Discussiones Mathematicae, Differential Inclusions, Control and Optimization, Vol. 2, pp. 209-244.

[KuK21] Kundu, S., and Kunisch, K., 2021. "Policy Iteration for Hamilton-Jacobi-Bellman Equations with Control Constraints," Computational Optimization and Applications, pp. 1-25.

[KuV86] Kumar, P. R., and Varaiya, P. P., 1986. Stochastic Systems: Estimation, Identification, and Adaptive Control, Prentice-Hall, Englewood Cliffs, N. J.

[KuL82] Kumar, P. R., and Lin, W., 1982. "Optimal Adaptive Controllers for Unknown Markov Chains," IEEE Trans. Automatic Control, Vol. AC-27, pp. 765-774.

[Kum83] Kumar, P. R., 1983. "Optimal Adaptive Control of Linear-Quadratic-Gaussian Systems," SIAM J. on Control and Optimization, Vol. 21, pp. 163-178.

[Kum85] Kumar, P. R., 1985. "A Survey of Some Results in Stochastic Adaptive Control," SIAM J. on Control and Optimization, Vol. 23, pp. 329-380.

[LAM21] Lopez, V. G., Alsalti, M., and Muller, M. A., 2021. "Efficient Off-Policy Q-Learning for Data-Based Discrete-Time LQR Problems," arXiv preprint arXiv:2105.07761.

[LJM21] Li, Y., Johansson, K. H., Martensson, J., and Bertsekas, D. P., 2021. "Data-Driven Rollout for Deterministic Optimal Control," arXiv preprint arXiv:-2105.03116.

[LLL08] Lewis, F. L., Liu, D., and Lendaris, G. G., 2008. Special Issue on Adaptive Dynamic Programming and Reinforcement Learning in Feedback Control, IEEE Trans. on Systems, Man, and Cybernetics, Part B, Vol. 38, No. 4.

[LPS21] Liu, M., Pedrielli, G., Sulc, P., Poppleton, E., and Bertsekas, D. P., 2021. "ExpertRNA: A New Framework for RNA Structure Prediction," bioRxiv 2021.01.18.427087; to appear in INFORMS J. on Computing.

[LWW17] Liu, D., Wei, Q., Wang, D., Yang, X., and Li, H., 2017. Adaptive Dynamic Programming with Applications in Optimal Control, Springer, Berlin.

[LXZ21] Liu, D., Xue, S., Zhao, B., Luo, B., and Wei, Q., 2021. "Adaptive Dynamic Programming for Control: A Survey and Recent Advances," IEEE Transactions on Systems, Man, and Cybernetics, Vol. 51, pp. 142-160.

[LaR95] Lancaster, P., and Rodman, L., 1995. Algebraic Riccati Equations, Clarendon Press.

[LaS20] Lattimore, T., and Szepesvari, C., 2020. Bandit Algorithms, Cambridge Univ. Press.

[LaW13] Lavretsky, E., and Wise, K., 2013. Robust and Adaptive Control with Aerospace Applications, Springer.

[LeL13] Lewis, F. L., and Liu, D., (Eds), 2013. Reinforcement Learning and Approximate Dynamic Programming for Feedback Control, Wiley, Hoboken, N. J.

[LeV09] Lewis, F. L., and Vrabie, D., 2009. "Reinforcement Learning and Adaptive Dynamic Programming for Feedback Control," IEEE Circuits and Systems Magazine, 3rd Q. Issue.

[Li17] Li, Y., 2017. "Deep Reinforcement Learning: An Overview," arXiv preprint ArXiv: 1701.07274v5.

[MDM01] Magni, L., De Nicolao, G., Magnani, L., and Scattolini, R., 2001. "A Stabilizing Model-Based Predictive Control Algorithm for Nonlinear Systems," Automatica, Vol. 37, pp. 1351-1362.

[MKH10] Mayer, J., Khairy, K., and Howard, J., 2010. "Drawing an Elephant with Four Complex Parameters," American Journal of Physics, Vol. 78, pp. 648-649.

[MRR00] Mayne, D., Rawlings, J. B., Rao, C. V., and Scokaert, P. O. M., 2000. "Constrained Model Predictive Control: Stability and Optimality," Automatica,

Vol. 36, pp. 789-814.

[MVB20] Magirou, E. F., Vassalos, P., and Barakitis, N., 2020. "A Policy Iteration Algorithm for the American Put Option and Free Boundary Control Problems," J. of Computational and Applied Mathematics, vol. 373, p. 112544.

[MaK12] Mausam, and Kolobov, A., 2012. "Planning with Markov Decision Processes: An AI Perspective," Synthesis Lectures on Artificial Intelligence and Machine Learning, Vol. 6, pp. 1-210.

[Mac02] Maciejowski, J. M., 2002. Predictive Control with Constraints, Addison-Wesley, Reading, MA.

[Man74] Mandl, P., 1974. "Estimation and Control in Markov Chains," Advances in Applied Probability, Vol. 6, pp. 40-60.

[May14] Mayne, D. Q., 2014. "Model Predictive Control: Recent Developments and Future Promise," Automatica, Vol. 50, pp. 2967-2986.

[Mes16] Mesbah, A., 2016. Stochastic Model Predictive Control: An Overview and Perspectives for Future Research," IEEE Control Systems Magazine, Vol. 36, pp. 30-44.

[Mey07] Meyn, S., 2007. Control Techniques for Complex Networks, Cambridge Univ. Press, N. Y.

[NaA12] Narendra, K. S., and Annaswamy, A. M., 2012. Stable Adaptive Systems, Courier Corporation.

[OrR67] Ortega, J. M., and Rheinboldt, W. C., 1967. "Monotone Iterations for Nonlinear Equations with Application to Gauss-Seidel Methods," SIAM J. on Numerical Analysis, Vol. 4, pp. 171-190.

[OrR70] Ortega, J. M., and Rheinboldt, W. C., 1970. Iterative Solution of Nonlinear Equations in Several Variables, Academic Press; republished in 2000 by the Society for Industrial and Applied Mathematics.

[PaJ21] Pang, B., and Jiang, Z. P., 2021. "Robust Reinforcement Learning: A Case Study in Linear Quadratic Regulation," arXiv preprint arXiv:2008.11592v3.

[Pan90] Pang, J. S., 1990. "Newton's Method for B-Differentiable Equations," Math. of Operations Research, Vol. 15, pp. 311-341.

[PoA69] Pollatschek, M., and Avi-Itzhak, B., 1969. "Algorithms for Stochastic Games with Geometrical Interpretation," Management Science, Vol. 15, pp. 399-413.

[PoR12] Powell, W. B., and Ryzhov, I. O., 2012. Optimal Learning, J. Wiley, N. Y.

[PoV04] Powell, W. B., and Van Roy, B., 2004. "Approximate Dynamic Programming for High-Dimensional Resource Allocation Problems," in Learning and Approximate Dynamic Programming, by J. Si, A. Barto, W. Powell, and D. Wunsch, (Eds.), IEEE Press, N. Y.

[Pow11] Powell, W. B., 2011. Approximate Dynamic Programming: Solving the Curses of Dimensionality, 2nd Edition, J. Wiley and Sons, Hoboken, N. J.

[PuB78] Puterman, M. L., and Brumelle, S. L., 1978. "The Analytic Theory of Policy Iteration," in Dynamic Programming and Its Applications, M. L. Puterman (ed.), Academic Press, N. Y.

[PuB79] Puterman, M. L., and Brumelle, S. L., 1979. "On the Convergence of

Policy Iteration in Stationary Dynamic Programming," Math. of Operations Research, Vol. 4, pp. 60-69.

[Put94] Puterman, M. L., 1994. Markovian Decision Problems, J. Wiley, N. Y.

[Qi93] Qi, L., 1993. "Convergence Analysis of Some Algorithms for Solving Nonsmooth Equations," Math. of Operations Research, Vol. 18, pp. 227-244.

[QiS93] Qi, L., and Sun, J., 1993. "A Nonsmooth Version of Newton's Method," Math. Programming, Vol. 58, pp. 353-367.

[RMD17] Rawlings, J. B., Mayne, D. Q., and Diehl, M. M., 2017. Model Predictive Control: Theory, Computation, and Design, 2nd Ed., Nob Hill Publishing (updated in 2019 and 2020).

[Rec18] Recht, B., 2018. "A Tour of Reinforcement Learning: The View from Continuous Control," Annual Review of Control, Robotics, and Autonomous Systems.

[RoB17] Rosolia, U., and Borrelli, F., 2017. "Learning Model Predictive Control for Iterative Tasks. A Data-Driven Control Framework," IEEE Trans. on Automatic Control, Vol. 63, pp. 1883-1896.

[RoB19] Rosolia, U., and Borrelli, F., 2019. "Sample-Based Learning Model Predictive Control for Linear Uncertain Systems," 58th Conference on Decision and Control (CDC), pp. 2702-2707.

[Rob80] Robinson, S. M., 1980. "Strongly Regular Generalized Equations," Math. of Operations Research, Vol. 5, pp. 43-62.

[Rob88] Robinson, S. M., 1988. "Newton's Method for a Class of Nonsmooth Functions," Industrial Engineering Working Paper, University of Wisconsin; also in Set-Valued Analysis Vol. 2, 1994, pp. 291-305.

[Rob11] Robinson, S. M., 2011. "A Point-of-Attraction Result for Newton's Method with Point-Based Approximations," Optimization, Vol. 60, pp. 89-99.

[SGG15] Scherrer, B., Ghavamzadeh, M., Gabillon, V., Lesner, B., and Geist, M., 2015. "Approximate Modified Policy Iteration and its Application to the Game of Tetris," J. of Machine Learning Research, Vol. 16, pp. 1629-1676.

[SBP04] Si, J., Barto, A., Powell, W., and Wunsch, D., (Eds.) 2004. Learning and Approximate Dynamic Programming, IEEE Press, N. Y.

[SHM16] Silver, D., Huang, A., Maddison, C. J., Guez, A., Sifre, L., Van Den Driessche, G., Schrittwieser, J., Antonoglou, I., Panneershelvam, V., Lanctot, M., and Dieleman, S., 2016. "Mastering the Game of Go with Deep Neural Networks and Tree Search," Nature, Vol. 529, pp. 484-489.

[SHS17] Silver, D., Hubert, T., Schrittwieser, J., Antonoglou, I., Lai, M., Guez, A., Lanctot, M., Sifre, L., Kumaran, D., Graepel, T., and Lillicrap, T., 2017. "Mastering Chess and Shogi by Self-Play with a General Reinforcement Learning Algorithm," arXiv preprint arXiv:1712.01815.

[SSS17] Silver, D., Schrittwieser, J., Simonyan, K., Antonoglou, I., Huang, A., Guez, A., Hubert, T., Baker, L., Lai, M., Bolton, A., and Chen, Y., 2017. "Mastering the Game of Go Without Human Knowledge," Nature, Vol. 550, pp. 354-359.

[SYL04] Si, J., Yang, L., and Liu, D., 2004. "Direct Neural Dynamic Programming," in Learning and Approximate Dynamic Programming, by J. Si, A. Barto, W. Powell, and D. Wunsch, (Eds.), IEEE Press, N. Y.

[SaB11] Sastry, S., and Bodson, M., 2011. Adaptive Control: Stability, Convergence and Robustness, Courier Corporation.

[SaL79] Saridis, G. N., and Lee, C.-S. G., 1979. "An Approximation Theory of Optimal Control for Trainable Manipulators," IEEE Trans. Syst., Man, Cybernetics, Vol. 9, pp. 152-159.

[SaR04] Santos, M. S., and Rust, J., 2004. "Convergence Properties of Policy Iteration," SIAM J. on Control and Optimization, Vol. 42, pp. 2094-2115.

[Sch15] Schmidhuber, J., 2015. "Deep Learning in Neural Networks: An Overview," Neural Networks, pp. 85-117.

[Sha53] Shapley, L. S., 1953. "Stochastic Games," Proc. of the National Academy of Sciences, Vol. 39, pp. 1095-1100.

[SlL91] Slotine, J.-J. E., and Li, W., Applied Nonlinear Control, Prentice-Hall, Englewood Cliffs, N. J.

[Str66] Strauch, R., 1966. "Negative Dynamic Programming," Ann. Math. Statist., Vol. 37, pp. 871-890.

[SuB18] Sutton, R., and Barto, A. G., 2018. Reinforcement Learning, 2nd Ed., MIT Press, Cambridge, MA.

[Sze10] Szepesvari, C., 2010. Algorithms for Reinforcement Learning, Morgan and Claypool Publishers, San Franscisco, CA.

[TeG96] Tesauro, G., and Galperin, G. R., 1996. "On-Line Policy Improvement Using Monte Carlo Search," NIPS, Denver, CO.

[Tes94] Tesauro, G. J., 1994. "TD-Gammon, a Self-Teaching Backgammon Program, Achieves Master-Level Play," Neural Computation, Vol. 6, pp. 215-219.

[Tes95] Tesauro, G. J., 1995. "Temporal Difference Learning and TD-Gammon," Communications of the ACM, Vol. 38, pp. 58-68.

[TsV96] Tsitsiklis, J. N., and Van Roy, B., 1996. "Feature-Based Methods for Large-Scale Dynamic Programming," Machine Learning, Vol. 22, pp. 59-94.

[VVL13] Vrabie, D., Vamvoudakis, K. G., and Lewis, F. L., 2013. Optimal Adaptive Control and Differential Games by Reinforcement Learning Principles, The Institution of Engineering and Technology, London.

[Van67] Vandergraft, J. S., 1967. "Newton's Method for Convex Operators in Partially Ordered Spaces," SIAM J. on Numerical Analysis, Vol. 4, pp. 406-432.

[Van78] van der Wal, J., 1978. "Discounted Markov Games: Generalized Policy Iteration Method," J. of Optimization Theory and Applications, Vol. 25, pp. 125-138.

[WLL16] Wei, Q., Liu, D., and Lin, H., 2016. "Value Iteration Adaptive Dynamic Programming for Optimal Control of Discrete-Time Nonlinear Systems," IEEE Transactions on Cybernetics, Vol. 46, pp. 840-853.

[WLL21] Winnicki, A., Lubars, J., Livesay, M., and Srikant, R., 2021. "The Role of Lookahead and Approximate Policy Evaluation in Policy Iteration with Linear Value Function Approximation," arXiv preprint arXiv:2109.13419.

[WhS92] White, D., and Sofge, D., (Eds.), 1992. Handbook of Intelligent Control, Van Nostrand, N. Y.

[YuB13] Yu, H., and Bertsekas, D. P., 2013. "Q-Learning and Policy Iteration Algorithms for Stochastic Shortest Path Problems," Annals of Operations Research,

Vol. 208, pp. 95-132.

[YuB15] Yu, H., and Bertsekas, D. P., 2015. "A Mixed Value and Policy Iteration Method for Stochastic Control with Universally Measurable Policies," Math. of OR, Vol. 40, pp. 926-968.

[ZSG20] Zoppoli, R., Sanguineti, M., Gnecco, G., and Parisini, T., 2020. Neural Approximations for Optimal Control and Decision, Springer.

Neuro-Dynamic Programming
Dimitri P. Bertsekas and John N. Tsitsiklis

Athena Scientific, 1996
512 pp., hardcover, ISBN 1-886529-10-8

This is the first textbook that fully explains the neuro-dynamic programming/reinforcement learning methodology, a breakthrough in the practical application of neural networks and dynamic programming to complex problems of planning, optimal decision making, and intelligent control.

From the review by George Cybenko for IEEE Computational Science and Engineering, May 1998:

"Neuro-Dynamic Programming is a remarkable monograph that integrates a sweeping mathematical and computational landscape into a coherent body of rigorous knowledge. The topics are current, the writing is clear and to the point, the examples are comprehensive and the historical notes and comments are scholarly."

"In this monograph, Bertsekas and Tsitsiklis have performed a Herculean task that will be studied and appreciated by generations to come. I strongly recommend it to scientists and engineers eager to seriously understand the mathematics and computations behind modern behavioral machine learning."

Among its special features, the book:

- Describes and unifies a large number of NDP methods, including several that are new

- Describes new approaches to formulation and solution of important problems in stochastic optimal control, sequential decision making, and discrete optimization

- Rigorously explains the mathematical principles behind NDP

- Illustrates through examples and case studies the practical application of NDP to complex problems from optimal resource allocation, optimal feedback control, data communications, game playing, and combinatorial optimization

- Presents extensive background and new research material on dynamic programming and neural network training

Neuro-Dynamic Programming is the winner of the 1997 INFORMS CSTS prize for research excellence in the interface between Operations Research and Computer Science

> **Reinforcement Learning and Optimal Control**
> Dimitri P. Bertsekas
>
> Athena Scientific, 2019
> 388 pp., hardcover, ISBN 978-1-886529-39-7

This book explores the common boundary between optimal control and artificial intelligence, as it relates to reinforcement learning and simulation-based neural network methods. These are popular fields with many applications, which can provide approximate solutions to challenging sequential decision problems and large-scale dynamic programming (DP). The aim of the book is to organize coherently the broad mosaic of methods in these fields, which have a solid analytical and logical foundation, and have also proved successful in practice.

The book discusses both approximation in value space and approximation in policy space. It adopts a gradual expository approach, which proceeds along four directions:

- From exact DP to approximate DP: We first discuss exact DP algorithms, explain why they may be difficult to implement, and then use them as the basis for approximations.

- From finite horizon to infinite horizon problems: We first discuss finite horizon exact and approximate DP methodologies, which are intuitive and mathematically simple, and then progress to infinite horizon problems.

- From model-based to model-free implementations: We first discuss model-based implementations, and then we identify schemes that can be appropriately modified to work with a simulator.

The mathematical style of this book is somewhat different from the one of the author's DP books, and the 1996 neuro-dynamic programming (NDP) research monograph, written jointly with John Tsitsiklis. While we provide a rigorous, albeit short, mathematical account of the theory of finite and infinite horizon DP, and some fundamental approximation methods, we rely more on intuitive explanations and less on proof-based insights. Moreover, our mathematical requirements are quite modest: calculus, a minimal use of matrix-vector algebra, and elementary probability (mathematically complicated arguments involving laws of large numbers and stochastic convergence are bypassed in favor of intuitive explanations).

The book is supported by on-line video lectures and slides, as well as new research material, some of which has been covered in the present monograph.

> **Rollout, Policy Iteration, and Distributed
> Reinforcement Learning**
>
> Dimitri P. Bertsekas
>
> Athena Scientific, 2020
> 480 pp., hardcover, ISBN 978-1-886529-07-6

This book develops in greater depth some of the methods from the author's Reinforcement Learning and Optimal Control textbook (Athena Scientific, 2019). It presents new research, relating to rollout algorithms, policy iteration, multiagent systems, partitioned architectures, and distributed asynchronous computation.

The application of the methodology to challenging discrete optimization problems, such as routing, scheduling, assignment, and mixed integer programming, including the use of neural network approximations within these contexts, is also discussed.

Much of the new research is inspired by the remarkable AlphaZero chess program, where policy iteration, value and policy networks, approximate lookahead minimization, and parallel computation all play an important role.

Among its special features, the book:

- Presents new research relating to distributed asynchronous computation, partitioned architectures, and multiagent systems, with application to challenging large scale optimization problems, such as combinatorial/discrete optimization, as well as partially observed Markov decision problems.

- Describes variants of rollout and policy iteration for problems with a multiagent structure, which allow the dramatic reduction of the computational requirements for lookahead minimization.

- Establishes connections of rollout algorithms and model predictive control, one of the most prominent control system design methodology.

- Expands the coverage of some research areas discussed in the author's 2019 textbook Reinforcement Learning and Optimal Control.

- Provides the mathematical analysis that supports the Newton step interpretations and the conclusions of the present book.

The book is supported by on-line video lectures and slides, as well as new research material, some of which has been covered in the present monograph.